AUTO
BIOGRAPHY

MAT WATSON

AUTO BIOGRAPHY

An A–Z Exposé of Cars

C

CENTURY

CENTURY

UK | USA | Canada | Ireland | Australia
India | New Zealand | South Africa

Century is part of the Penguin Random House group of companies
whose addresses can be found at global.penguinrandomhouse.com

Penguin Random House UK,
One Embassy Gardens, 8 Viaduct Gardens, London SW11 7BW

penguin.co.uk
global.penguinrandomhouse.com

Penguin
Random House
UK

First published 2025

001

Set in Calluna 11.7/16pt
Typeset by Six Red Marbles UK, Thetford, Norfolk
Printed and bound in India by Thomson Press India Ltd.

The authorised representative in the EEA is Penguin Random House Ireland,
Morrison Chambers, 32 Nassau Street, Dublin D02 YH68

A CIP catalogue record for this book is available from the British Library

ISBN: 978-1-529-96698-5

For my amazing daughter, Grace, because I always thought it would be cool to have a book dedicated to me. So now you do.

Contents

CONTENTS

CONTENTS

Introduction

As I'm sitting here writing this, I'm slightly distracted by the sub-scriber count on Carwow's YouTube page. We've only got eighty subscribers to go to hit 10 million. I remember back in 2017 when we were about to reach the 100,000 subscribers milestone, and for some reason I was a little apprehensive. Would we make it? Then, some of the coders in a corner of the office thought it'd be amusing to subscribe and then unsubscribe, so when the number count finally reached 100,000, it shot back down to 99,997. It was pretty funny and very in keeping with the tone of what we do. But it all feels different this time. Nothing's going to stop us motoring over the line.

Could I have imagined this as a nine-year-old from Walsall in a remedial class at primary school because the teachers thought I was dim? No. I remember a couple of other kids in that class – there was Lindsay, who used to eat her scabs. And then there was Ivan, who kept his dinner money in his mouth because he was worried that other kids would take it. I wonder what they're both doing now. Who knows: maybe Lindsay's a professor of haema-tology and Ivan's a millionaire hedge fund manager.

My senior school was a very average comprehensive. I did enjoy maths and science because I liked figuring out how things worked – I'd see a gadget and want to take it apart to see which components did what. I was in the lowest of the low classes for English literature, mainly because I couldn't sit still long enough to read entire books. Yet, I ended up with an A for GCSE because it was coursework-based and my mum did most of it. For the exam part, we were handed a poem to analyse and we were given several hours spread over two days in a classroom, under controlled conditions. Everyone was trying to read the poem and work out what the hell's going on. Meanwhile, I rewrote the poem on a separate piece of paper, took it home and asked my mum to analyse it. I wrote down what she said and smuggled it in the next day. The English language GCSE, however, was exam-based so no one was helping me out in that room, but I got an A. Looking back, perhaps I could have got an A in English literature on my own had I put the work in. Maybe, in a way, I cheated myself.

Because I struggled academically at school – I found it very hard to concentrate – I explored creative ways to achieve what I needed to. That often involved taking the odd calculated risk, occasional mischief and a fair amount of blagging. And it wasn't just a case of the ends justifying the means, because the means were entertaining. And that attitude has helped steer me to where I am today. It turns out that taking risks, being silly and blagging equips you well as a motoring journalist on YouTube.

In the 1980s, you were told that you needed to study the sciences if you were going to get a job. My sister had also gone down that route, taking the three sciences at A level, so that encouraged me along the same lines. She always knew what she wanted to do, aced her A levels, got into Cambridge University and became a vet. It took the pressure off me, because she'd done really well, so anything I brought to the table was a bonus. My teachers thought I'd fail my A levels because I wasn't able to concentrate and messed about too much, so to spite them I bought a bunch of textbooks and memorised them. I ended up with the fourth-highest grades in my year and got into the University of Edinburgh.

I applied to do chemical engineering largely because I met a chemical engineer who was making a lot of money. But I soon realised that the maths was beyond me so I switched to straight chemistry, which, annoyingly, seemed to become maths towards the final year. I didn't have the patience to become a chemist. But I did have my resourceful, creative streak to call upon. I remember a big test involving 'unknown' chemicals, and it was our job to go away and deduce what they might be. What that means is there's a label on a bottle, which they've struck through with a permanent marker so you can't see it. So, rather than conduct a sequence of experiments to find out what the chemical was, I got to work finding a solvent that would remove the permanent marker pen but preserve the label. I was still using chemistry to get the job done but it wasn't quite the problem-solving they were after!

In my final year at uni, recruiters put on these employment fairs to show you what their business is about. My dad was encouraging me towards accounting because it'd give me a good grounding in business and would pay pretty well, so I applied for a job at PricewaterhouseCoopers (PwC). But something didn't feel right: I was bored shitless. As soon as I qualified, I handed in my notice. I was into motorbikes back then, possibly to liven up the extremely dull work I was doing in the daytime, but then I had an accident and figured I'd be safer shifting my interest to cars. One of my mates at PwC had a girlfriend who was working as an editor on a local evening newspaper. She was the only person I'd ever met who spoke with passion about what they did. She genuinely loved her job, and that set something off in me. I realised that I was quite good at asking questions, finding stuff out and telling a story. So I ditched the golden handshake of a £45k role at a consulting firm for a job at a local newspaper. 'What on earth do you think you're doing?!' said my dad when I told him the salary. It was £12k.

After working at the local paper for a year, a friend of mine who knew how into cars I was showed me a job advert for a consumer writer on *Auto Express* magazine. I applied for it and got it partly because I'd studied business law as part of my accountancy training. This meant that when I was negotiating with car manufacturers about consumer rights, I was coming at it from a legal position. It turned out that I was really good at getting people new cars under the Sale and Supply of Goods Act 1994. Even from the get-go, I was always consumer-focused.

That job paid about £20k, which to be fair was some uplift from the job on the local paper, but the money didn't matter. As a kid growing up in Walsall in the Midlands, I'd fallen in love with cars the moment I saw my mate's dad's Jaguar XJS pull into the drive. What a thing of beauty that was, even in the yellowy beige paint job that he'd gone for. No one was thinking about residual values back then – they went for whichever colour they actually wanted. It was a simpler time. Look in a car park these days and you'll see fifty shades of grey.

When I was sixteen, my next-door neighbour bought himself a Triumph Spitfire – man, I loved that car. All I wanted was to pass my driving test as soon as I turned seventeen, and seeing as I'm a July baby, I was behind the curve already compared to my schoolmates. So my neighbour kindly took me out around the local trading estate out of hours so I could get used to the car. I didn't want to put a scratch on it, so I used my dad's car for the grunt work of actually learning. Things like hill starts, emergency stops and anything else with a likelihood that I'd prang it, I'd be in my dad's Mini Metro. We're talking a 1.0-litre engine with acceleration similar to those street-sweeper vans owned by the council. The interior was, let's say, 'minimalist' and I've used napkins that were thicker than the bodywork.

My mum and dad each had a car, and slightly unusually for

the time, her one was a lot cooler than my dad's. Hers was a Ford Fiesta XR2i and even though the term hadn't been used yet at that point in the late 1980s, that was a hot hatch. Eight-valve, fuel-injected 104hp engine, with alloy wheels and body stripes – that was the car you wanted at seventeen. Sure, we all dreamed of the supercars, but the XR2i felt like an achievable aspiration. Not that I got to drive it that much. I was in the Mini Metro: the first crap car I became fond of.

Being a proud Scot, my dad was keen on saving a few quid wherever possible, so he gave me a few casual lessons in the Metro rather than having to shell out for an instructor. But that worked well for me, because we'd spend quite a bit of time together at the weekends. The car was the kind of environment where fathers and sons can bond, because you had a distraction to take the pressure off any kind of serious conversation that suddenly sideswiped you. It's the reason why fathers and sons often have surprisingly deep conversations at the football. The focus is elsewhere.

But when the date of my driving test came closer, my parents organised four official lessons to teach me what I needed to know. Three months after my seventeenth birthday, I passed my test – first time – but that goes without saying. If I'd failed I wouldn't be able to live that one down, doing what I do! And when I did pass, suddenly that Mini Metro was the best car on the planet, because it gave me something no other car has ever given me: that first taste of freedom. Everyone feels the same when you get into a car on your own after passing your test. You've spent weeks or months with someone else next to you telling you what to do and where to go. And then it's just you, and you can go wherever you want. It's one of most electrifying moments of your life. Only, I didn't get to enjoy it for that long because I wrote it off within a month.

The silver lining was that it meant that I got to drive my mum's Ford Fiesta XR2i. After you'd driven a Mini Metro, the XR2i was like a Bugatti. I was driving it around literally everywhere for about a year, at which point my dad decided that he'd prefer us not to be driving around in a fast car, so they bought a second-hand Fiesta Mk1 for me and my sister to 'share'. My sister didn't really drive it a lot, though, so I ended up 'acquiring' it. Presents to share somehow ended up in my room during our adolescence. Sorry, sis! 'Our' Fiesta was canary yellow, except for the primer grey door that announced that the previous owner had pranged it but couldn't be arsed to pay for the spray job. That car – the Banana – took me everywhere. I hammered up the M6, fully loaded with stuff, towards my university halls in Edinburgh, including Keith, my royal python, in a tote bag tied up with a shoelace, curled up happily on the front seat.

As a kid, I was fascinated by snakes and I always wanted one. So, when I was eighteen and had a bit of cash, I bought one. Royal pythons are docile and placid, so it seemed like a good entry-level snake. I wasn't prepared for quite how docile Keith was, though, remaining motionless for weeks at a time. There's a low-maintenance animal and then there's having a rock for a pet. When I tried to handle him, he'd just slither away into a dark corner. People were fascinated by Keith at university, though. The trouble was, he started rejecting dead mice so I had to get hold of live ones. But Keith wouldn't go for the live ones either, so I found myself with a whole bunch of pet mice. I ended up liking the rodents more than the snake! Some of my friends and family members took them in as well, forming a kind of rejected rodents club. All the while, there Keith lay, doing almost nothing for many years. And then I returned from a holiday in 2015 and got the surprise of my life. Keith was sitting on top of several eggs. It turned out Keith was female and, apparently, some snakes are capable of reproducing asexually. So I renamed her Penelope Keith.

In my Ford Fiesta Mk1 (with curled-up Keith beside me), you had to seriously plan an overtake, a bit like an HGV driver. You see that downhill coming and you floor it, blazing past as many people as you can, riding the momentum and then dropping back between the HGVs. That was the full Formula 1 experience to a nineteen-year-old. I did some silly stuff in the car – we're all guilty of that in our first car – including giving in to the temptation to yank up the handbrake on a country lane. The car slid to the left, then swung to the right (hedgerow-assisted) and then came back to the centre. Everything was back to normal in a few seconds, and then the adrenaline kicked in. It made me want to experiment with the handbrake, and when you do, you won't stop until you've nailed a handbrake turn. It's one thing you don't tend to experience any more, because almost everything's got an electronic parking brake, except for a few cars like the Toyota GR Yaris, which I've got. The GR Yaris is essentially a rally car so it's designed to do handbrake turns. So when you pull up the handbrake, it disconnects the rear axle, which means you can rotate the car easier. You can even buy the kit to relocate the handbrake from the normal position to near the dash, like a rally car, where it's easier to pull. My kit is on the way from Japan now!

You were a lot more connected to the car you were driving back in the early nineties compared to nowadays. Everything needed pre-planning and a lot more energy, from the no-power steering to the wind-up windows and choke. The winter was a genuine problem. Sometimes the car wouldn't start, and you'd have to figure out why. More often than not it was the points (an electrical switch that makes and breaks the ignition system circuit), which really didn't like condensation. The starter motor on the Banana had a couple of teeth missing, which wasn't a problem unless the car had stopped with the teeth in the wrong position, because then the starter motor wouldn't turn the engine. It's usually no big deal, though, unless you've accidentally pissed off a really big guy who's chased you to your car and you're trying to get out of there pronto (true story).

All in all, the Banana served me really well. I didn't have any major problems with it. It was cheap to run, and gave me the ability to go wherever I wanted, whenever I wanted. It completely changed my life. We had it for just over two years and I sold it for the same money my parents paid for it, which felt like a win. Believe it or not, though, I'm not very good at selling cars. I seem to be better at buying them – except for the first-generation Mazda MX-5 I bought in 1997. I feel like I've been making up for that shambles of a transaction for the last twenty-five years.

In the last nine years, I've carved myself a niche at Carwow. From the moment I started shooting content for the Carwow YouTube channel, I tried to deliver a different type of video: something accessible, engaging and informed, with a good sprinkling of silliness. It helps that I'm not really beholden to anyone – apart from the viewers I'm reviewing the cars for. I think perhaps I have a freer rein than other motoring journalists because Carwow understand the value of content. It means that I can say whatever I want about whichever car I'm test driving. Car manufacturers never know what I'm going to say. But I do have to tread a fine line between not pissing off every single car manufacturer that is prepared to lend me a car, while producing content that's unique and different enough that people are going to want to watch it. And yes, sometimes that lands me in hot water, like with Mazda in 2020, when I said the Mazda 3 looked like a cat having a poo. Mazda still haven't forgiven me. I don't follow the textbook because I didn't read the textbook. I just do what comes naturally to me. That sometimes gets me into trouble, but I've got thick skin.

After a career spent kicking tyres, climbing into boots and drag-racing some of the world's fastest and scariest cars, it felt like the right time to share my knowledge and stories in a book. In my videos, I try to be impartial and think about cars from a potential buyer's point of view. Sure – there's always going to be a little subconscious bias towards what you think looks good, what you value in a car and, ultimately, what you'd buy yourself.

Some of the manufacturers I love; some of them I really don't; and some have especially memorable stories attached to them. You'll find out the answers to the questions: why am I such a Porsche fan boy? How did I manage to wind up Renault? Why is it that I like BMWs and yet I've only owned one of them? Why does everyone love a Ford Focus? Which hypercar would I go for? What's on my driveway? Why do I love a crapheap? Along the way, I'm lifting up the bonnet so you can discover what my life is really like as a motoring journalist.

Alfa Romeo

This whole hoopla about not being a car fanatic unless you love Alfa Romeos feels like it's been driven by Jeremy Clarkson, who's a big Alfa fan. I'm not an Alfa fan and I never have been. Yes, their cars have personality, a really good history and legacy, but if you like German cars you won't like them. And I like German cars.

I've driven lots of Alfa Romeos and have liked some, but they've just felt like also-rans for as long as I've been into cars. Maybe if I'd have got into cars twenty years earlier, in the 1960s, I would have felt differently – when the Alfa Spider came out in 1966. Or if I'd been around in 1923 when Enzo Ferrari was winning grands prix in Alfa Romeos. In 1929, Enzo founded the racing team Scuderia Ferrari but all the Scuderia Ferrari drivers were racing Alfa Romeos. The first Ferrari car was a long way off. And the first car to feature the prancing horse badge wasn't a Ferrari – it was the 1935 Alfa Romeo Bimotore, with the Ferrari logo painted on to the radiator cowl. Scuderia Ferrari was absorbed into Alfa Romeo in 1937, so Enzo Ferrari left in 1939 to found his own company the following year. Only he couldn't use the Ferrari name for the company or any of his cars for the next four years because he'd signed a non-compete clause with Alfa Romeo. Not that any of that mattered, given that most of Europe was at war.

The first Alfa Romeo I saw in the flesh was an Alfasud (produced from 1971–1983), owned by a guy across from us. It was a great little family car and a cool shape. The next one I really remember was the GTV6 (their sporty coupé, produced from 1993–2004) parked in a driveway down the road in the early 1990s. The GTV6 had an amazing interior with beautiful seats but it didn't handle great and suffered from torque-steer (where power in a front-wheel drive car is applied unevenly to the drive shafts so the steering pulls one way or another). It looked great in my neighbour's driveway, though, and I was able to get a good look at it seeing as it was constantly up on axle jacks in the drive while he fixed literally everything on the car. It was on those jacks more than it was on the road.

Alfa have struggled since the 1980s and their cars that have caught my attention feel like outliers. The Giulia Quadrifoglio came out in 2015 and was a really good car with an engine to match. They launched it at the Alfa Romeo museum in Milan while Italian tenor Andrea Bocelli sang Puccini's 'Nessun dorma', in case you'd forgotten that the company's Italian. The issue with the Giulia Quadrifoglio arises when you compare it to the competition – the BMW M4. The infotainment isn't as good on the Alfa, you couldn't turn the electric stability control (ESC) all the way off, so it's not as fun when you're on track, and it didn't have a dual-clutch automatic gearbox so it wasn't as responsive. But the engine was great and it did have some cool features like a super-light carbon-fibre prop shaft, which sends the power from the engine to the rear wheels very quickly, so the acceleration picks up really well (as does braking response). Plus, the car looks beautiful. But there were also little things that annoyed me, like the way they'd positioned the door – too far forward – which makes it harder to get in and makes you feel like the dash is too far back. It's things like that that make you want to play it safe and go with the BMW. And I think Alfa have suffered from that problem for decades. Specialised limited-edition versions of the Giulia Quadrifoglio, the GTA and GTAm, featured a lot more carbon-fibre and carbon-ceramic brake discs but prices started at £150k. That's not a typo. Yes, I am more likely to turn my head when I see the Giulia Quadrifoglio go down the road compared to an M4 but spending £150k on one? No. It's a nice car to look at and a nice car to borrow, but it's also a nice car to give back.

Their SUV, the Stelvio Quadrifoglio, is a similar story. It handles pretty well and it's got an amazing engine (as it's the same one as in the Giulia), but it's the same price as an equivalent BMW or Mercedes, and there are little bits about it where you can tell that less money has been spent. It's partly because the Quadrifoglio's rear-wheel drive platform is just designed for Alfa Romeo. This means that despite the fact that they're part of the Stellantis group (which also owns Peugeot, Citroën and Vauxhall among others), Alfa are not benefiting from the economies of scale achieved by sharing a platform, like say you can with a Peugeot 208, which shares a platform with the Vauxhall Corsa and Citroën C3. It means that Alfa's competitors for this car, like BMW and Mercedes, which both sell their rear-wheel-drive saloon and coupé models in much higher volumes, have had more cash to splash on their cars – not necessarily in terms of the engine, or even the interior fit and finish, but more in the development phase when they thought the whole car through. It's given them more time to check everything and make it as good as it can be.

Ultimately it all comes down to how much you've got to spend on developing a car. A good example of this is Hyundai. Recently, they've chucked money at developing their new 'N' line of high-performance hot hatches and it's paid off. Alfa just haven't got the cash.

Alfas have just been a bit two-dimensional. The issue is that their budget limitations mean that they never seem to create a fully rounded product. Either the looks or the engine (or sometimes both) was epic, but the whole package felt incomplete. The 3.2-litre direct-injection V6 engine on the Alfa 159 is a work of art, with its shiny intake pipes, and it sounds incredible but it's in a chassis that's just meh. The 147, with that 3.2-litre V6 engine, was good. Mad but good. Again, though, the trouble arrives when you start thinking about the competition. I'd rather have a Renault Sport Megane because it was so much better dynamically. It was all about the engine in the Alfa but you felt like it was trying to get away from you, a bit like you were attempting to restrain a crazy Italian stallion, full of sound and fury.

Alfa do make pretty things but the ergonomics inside often aren't great and the touchscreens are low-def. It's things like that. And there's the whole history of reliability problems. None of them have given me *that feeling* – when you sense the car is communicating with you while you're driving. Bits were good working in isolation, but they didn't come together as a team. A lot like Paris Saint-Germain when they had Neymar, Mbappé and Messi in the same team. Or England in the early noughties when we had Beckham, Ferdinand, Gerrard, Lampard, Scholes and Rooney. We should have won something but we didn't.

Here's a good example of Alfa dropping the ball. I remember seeing their 4C (mid-engined sports car, produced from 2013 as a coupé and 2015 as a Spider) when it was unveiled at the Geneva Motor Show in 2013. It had a 1.8-litre turbocharged petrol engine with a dual-clutch automatic gearbox and, looks-wise, it was a stunning, stunning car. Incredibly, it was carbon tubbed (i.e., the whole chassis was made of carbon-fibre) and no one, except a Formula 1 team or McLaren Automotive, makes a carbon-tubbed car. It was a decent price too, for a carbon-tubbed car, around the same as a Porsche Cayman. So far so good, right? Everyone was desperate for it to drive as beautifully as it looked. It didn't.

I don't know what they did to it, but the 4C managed to both understeer and oversteer. It was sketchy and, most importantly, it didn't feel fun. It's a bit odd when something that stunning isn't fun. I've thought about buying a 4C many times. But I'd never end up driving it. I'd just look at it.

The Alfa MiTo was a front-wheel drive supermini (and competitor to the new Mini) that came out in 2008. It was brilliant, with its little turbocharged engine, and I thought, *Here we go – people are going to buy these like they do the Mini*, but they just didn't. The Giulietta was a really pretty hatchback that Alfa produced between 2010 and 2020. The rear end on that was just stunning, but once again, when you get into the detail of reviewing the car as a whole package, it started to fall short. There was a period when you saw quite a few Giuliettas on the road but where have they all gone? It's like there's been some kind of Giulietta extinction event. It's weird. Whenever I did see one, I used to try to catch up with it and sit behind it, just to look at it. It's the only small family hatchback that I can think of where I would deliberately miss my junction on the motorway, adding another five minutes to the journey – which I'd never ordinarily do – just to keep looking at that car. I'm now going to consciously look out for Giuliettas to ascertain if I just haven't been registering them or if they've actually disappeared. Maybe they're up on axle jacks. Maybe they've just left my brain.

A bit like Fiat, Alfa Romeo's press launches back in the noughties summed up the company well. The whole event would take up three days of your time but you'd only be in the car for three hours. I'm not sure they wanted motoring journalists to really test the cars. One example was with the facelifted 166, which was billed as an executive saloon. This was the kind of car that you want to test by cruising on the motorway and then on a few twisty country roads because it's an Alfa and it's supposed to handle well. Well, they launched the facelifted 166 in the centre of either Berlin or Barcelona – I can't actually remember, but wherever it was, I spent the entire time driving it in stop-start traffic. Imagine a car manufacturer thinking it was a good idea to fly dozens of motoring journalists to the busiest stretch of the North Circular in London. After spending three hours in traffic, Alfa took us for dinner and we were all sat around these tables with one seat left between us and the next motoring journalist. It was really weird because we had starters, then main courses and no one had explained what was going on with the empty chair situation. Then a door opened and a procession of glamour models walked in and sat in each of the empty chairs. It was so awkward. It was like one of those boys' nights where you leave the club and are shepherded along by a chorus of encouraging 'waheys' towards a lap dancing club. It's tragic.

Until around 2012, 'stand girls' would be a regular furnishing feature at major motor shows to encourage punters towards particular manufacturers. We've all moved past this now, thankfully. Car manufacturers are still in the business of making sure motoring journalists are having a great time at a press event or a launch, though, because if you're teetering on the edge of a three- or four-star review, being in a good mood can make the all-important difference. And if someone's happy, they'll probably write more positive copy. In the noughties, one car manufacturer routinely put on UK launches of a car in a foreign country. They did the metrics on it, and realised that if the weather was good and the route they chose was along quiet, smooth and scenic roads they'd probably get a higher average mark from each journalist. Other manufacturers soon followed suit. So they'd fly you out on a two-day trip and on the first day, you'd travel, drive the car for a bit, then you'd have coffee, have a chat and sit down for a nice dinner. The next morning, there might be some event in the morning like clay pigeon shooting or quad biking. Then you drive the car some more before they fly you home. All in all, you've had a very pleasant couple of days and you've only got to put together an 800-word review. The internet ended all this though. With a need to get things online immediately and produce not only written but also video and social content, no one has the time to have fun on a car launch.

In the noughties, free gifts, known in the trade as 'the blag', were handed out to motoring journalists. You'd come back from the launch and other journalists would ask you not just whether or not the car was any good but also if the blag was too. They'd range in value and quality but would typically be electrical items, clothing, that sort of thing. I remember getting a pocket digital camera at a launch that was decent, but the car was less so and I gave it a bit of a kicking in my review. In fact, sometimes I think the gifts made the journalist even harsher on the car to sort of prove the point that they couldn't be bought.

I think the last blag I got was a pair of Bowers & Wilkins headphones at an Aston Martin launch but that linked to the stereo in the car. It wasn't just random free stuff. I remember hearing of one famous blag in the eighties or nineties where, instead of a goodie bag, the journalist got to keep the car. It happened because it would have cost the manufacturer so much to export the car back that they just decided to cut their losses. It was a cheap, crappy hatchback and that presented the journalist with an awkward dilemma. Should I refuse the gift and offend the manufacturer or be polite and drive this shed back home?

BEST CAR LOGOS OF ALL TIME

1. **Ferrari** – The Prancing Horse on the canary yellow background (the colour of the coat of arms of Enzo Ferrari's hometown of Modena) is undoubtedly the most famous car badge of all time. And for good reason.

2. **Mercedes** – In some ways it's even better than Ferrari's logo because the three-pointed star is so simple. I love the position of the emblem on the bonnet of older Mercedes, acting like a gunsight.

3. **Lamborghini** – I love the fact that Lamborghini's raging bull logo was born from Ferruccio Lamborghini's vendetta against Ferrari.

4. **Jaguar (the Growler)** – Jaguar had two or three logos but the Growler – the head-on design of the animal growling – was so cool, as was the name. Such a shame they ditched it with their recent rebrand. What a mistake (like the rest of the rebrand).

5. **Abarth** – What's cooler than having a scorpion on the front of your car? A great emblem.

6. **Alfa Romeo** – What is it with the Italians with their brilliant badges? A serpent swallowing a man is madness, but I love it.

7. **Corvette** – It's got the chequered flag on it. Nuff said. Does exactly what it says on the tin.

8. **Saab** – I love the old logo's aeroplane symbol; the red griffin wearing the golden crown was introduced in 1984 and it's cool. Much cooler than the griffin badge on another manufacturer . . .

9. **Proton** – The stylised tiger's head is such a cool badge, it's hard to believe it's the work of a car manufacturer that makes such dull cars.

10. **MG** – When I was young, MGs were the coolest cars in the world to me and I wanted one so badly. The stylised M and G lettering perfectly filling the octagonal space of the badge is such a cool piece of design.

Aston Martin

Aston Martin are a bit like Jaguar Land Rover. They're another classic British company that have produced some great stuff . . . and some rubbish. Aston have been badly managed and have struggled financially, somehow managing to survive seven bankruptcies. But as it happened, 007 was their lucky number.

Ian Fleming did Aston Martin a big favour by having James Bond choose an Aston over a 3.4-litre Jaguar to temporarily replace his Bentley in the novel *Goldfinger*. The Aston was a DB Mk III, but they sent producer Cubby Broccoli their latest model: the DB5. Not that it mattered, because after the world was introduced to Bond's silver DB5 with 'modifications' from Q Branch, it became *the* Bond car and one of the best-loved cars of all time. Back in the noughties, I worked with a well-known motoring journalist who bought one for normal car money. He sold it not so long ago (it would have been tax-free) and while he didn't reveal the figure to me, I suspect it went for well over half a million. Cars, like fine art, can be an incredible investment, but a painting's never going to generate the same G-forces as a car.

My relationship with Aston Martin is generally pretty friendly and I tend to like their cars. They've invited me to some fancy events and allowed me to drive some of their most exclusive cars. The Valkyrie launch in Bahrain ticked both of those boxes.

The Valkyrie is like a Formula 1 car – it's all aerodynamics and downforce wrapped up in stunning packaging. And that makes sense – it was designed by engineering legend Adrian Newey, who's picked up twelve constructors' championships in F1 and is now the technical director of the Aston Martin team. The Valkyrie has a naturally aspirated V12 engine made by Cosworth that revs up to 12,000rpm. It's a phenomenal car. It's also completely berserk. You have to wear earplugs when you're driving it but that just makes you feel even more like an F1 driver on a race track. Downsides (and I'm really stretching here): it's not easy to get in and out, but you don't want to get out once you've got in, so forget that. What else can I think of? Maybe not the best choice for a car seat?

When we were test-driving the Valkyrie on the launch, I wondered what the Aston Martin guys were going to do with this car that all the journalists had been knackering. 'Do you have to scrap it?' I asked. 'Oh no – we'll sell it. It's the one in the magazines and the videos that got all the views. It's got provenance.' It doesn't matter that it's an earlier version of the car that's been used. This is the one that kicked it all off, and to some super-rich folks, that makes it all the more desirable.

I flew to Abu Dhabi in February 2025 to drag-race a Valkyrie against a Mercedes-AMG ONE, but the AMG ONE broke. Then a little while later, almost in sympathy, the Valkyrie's gearbox threw up a fault. This is the kind of problem that car manufacturers experience with hypercars all the time, and it's not something people generally know about until they actually own one. With mass-produced cars, they're pretty much fully developed when they go on sale, but exclusive, limited-run cars are always subject to ongoing development. They're like beta models, which would be fine if you'd upgraded to the latest version of Microsoft Office, not bought a multimillion-pound hypercar. But then again, some owners like the attention you get when a team comes over to fix your car. And they like to feel part of the experience of developing a car. They feel special.

There is another factor to consider: Mercedes aren't going to go bust anytime soon, but Aston Martin, based on their history, might. And then what support are you going to get for your car? But then I'm thinking like a regular car buyer, not a billionaire. I suppose if you've got that much cash and your car goes wrong, you could hire the ex-Aston engineers to come and work for you personally. Or just jump into the next hypercar in your volcanic lair.

With the gearbox fault in the Valkyrie, you don't just head on down to the local Aston dealership. Aston Martin connect to the car online from their Gaydon HQ in Warwickshire, then chopper a team of engineers out to wherever the car is. It's a completely different world. But it makes sense: the problem that you've just encountered with their car might be the first time it's ever happened to this model, so they want to investigate it immediately and then run updates on all the other Valkyries. There's a kind of unspoken agreement that because hypercar manufacturers are operating at the cutting edge of technology, car owners come to understand that they're basically part of the R&D department.

It's not always the case with hypercars, though. Bugatti took a different approach with the Veyron. When Ferdinand Piëch was chairman at VW and they were developing the nearly 1,000hp Veyron, they put it through the same paces as they would have done with any new VW. And the upshot is that the Bugatti Veyron felt totally finished. It's a car that your mum could happily get in and start fine. It's comfortable, reasonably quiet and it won't break. And that's because it was built like a Volkswagen Golf. The drawback is that the Veyron lost money on every one it sold, but Piëch was looking at the bigger picture. He's created a halo car that lifted the entire VW group. And when they came to develop the Chiron, they weren't starting from scratch this time. It was an evolution of an existing product and they had economies of scale this time. And they could charge much more for the Chiron because everyone had seen how good the Veyron was. It was a masterclass in loss leading.

Aston Martin don't have the same purchasing power and economies of scale that VW-owned Bugatti do. You could see the difference that kind of backing gives a carmaker when you look at Lamborghini. Before they were taken over by VW, Lamborghinis were beautiful and crazy but pretty rubbish when it came to reliability. After the VW acquisition, Lamborghinis have become cars you can drive daily. Aston Martin never really had that kind of capability, until the early 1990s. It all started when Ford looked to develop a sporty Jaguar (who were also owned by Ford between 1990 and 2008) but didn't want to fork out a fortune. So Jaguar designed a newly styled XJS but, unexpectedly, they abandoned the idea. Car designer Ian Callum later explained that it was because Ford wanted Jaguar to develop a completely new car instead, which eventually became the XK8. Aston Martin picked up the pieces, and that car became the DB7. Underneath, it's a Jaguar XJS.

After that, Aston Martin's cars started getting better and better. The DB9 was beautiful and a great car to drive. Then came the high-performance version, the DBS, which is the one James Bond drives (and spectacularly crashes) in *Casino Royale*. Well, it was a prototype of the DBS, because the car was still some way off entering production when they shot the film in 2006. Then there was the 2005 V8 Vantage, which I refer to as 'the really pretty one'. Designed by the Danish-American Henrik Fisker, it became the bestselling Aston of all time. A couple of years later, they unveiled a concept car that had started as a V8

Vantage, only they substituted the AMII VI2 engine from the DBS, added a retractable rear wing and carbon-ceramic brakes. It was a V8 Vantage in beast mode, essentially. Dr Ulrich Bez, Aston Martin's CEO, invited the public to come up with a name for its grey-blue colour scheme. It's not always a great strategy, and I'm sure 'Car McCarface' probably topped the poll. But Aston are the kind of folks that favour quality over quantity, and so they went with the outstanding suggestion from an M. Watson, a thirty-something car journalist at *Auto Express*. The grey-blue colour of the concept car reminded me of a shark, and because it was a super-quick car, I thought of the fastest shark in the ocean. So I went with 'mako blue'. And that colour is still in their palette now.

The current Vantage (unveiled in November 2017) got a lot of stick for its grille, which does look like a shark that's about to bite you, but I really liked the design. They'd borrowed some bits from Mercedes-AMG, like the small matter of the engine, but also the switchgear and infotainment system. But Mercedes being Mercedes didn't want them to have the latest AMG engine, so it was detuned. They also didn't let Aston have the latest infotainment system – or even the one before that! So the Vantage ended up with the one before that, which isn't great when you're buying a GT car that's coming up against a new Porsche 911, with all the latest tech inside. There are lots of gadgets that are of no real use to a car buyer, but the infotainment system isn't one of them. It's vital. You want it to be easy to use and you want your phone to connect as easily and quickly as possible.

Aston responded to these concerns and have upgraded the engine as part of their 2024 facelift of the Vantage, so it's now more powerful than the AMGs. And the designer interior is a nice touch. They've also added their own infotainment system, which is easier to use and much, much better than it was but still not as good as Mercedes'. This is one of the problems you face as a small manufacturer because the costs involved in developing an infotainment system are massive and they won't have the same budget or economies of scale as one of the bigger guys. A modified version of the latest Vantage has been the F1 safety car (well, one of two, alongside the Mercedes-AMG GT R) since 2021 and it's lost the *Jaws* vibe from the grille. The F1 medical car, also known as 'the fastest ambulance in the world', is a modified version of Aston's first SUV, the DBX.

No one was quite sure what to expect when Aston joined the SUV party in 2020. The DBX, unlike almost all of its rivals, was designed from the ground up. While they don't have the platform-sharing ability that VW's brands do and so they have less money to develop the car, there are some advantages. In a platform-sharing arrangement, there are certain structural 'hard points' on the car, which have to remain the same. But Aston Martin don't have to worry about that so they have more freedom to tinker with the wheelbase (the distance between the front and rear axles), which can alter the overall shape of the car. As a result, the DBX has a longer wheelbase and therefore shorter overhangs. This gives it a sleeker design and makes it more aerodynamic. If you're after a performance SUV that's sporty to drive, the DBX is the best there is. It should sell better than it does, but British brands do have a reputation for reliability issues and catastrophic depreciation. In the mind of your high-end car buyer, those worries, whether they're founded or unfounded, are hard to shift.

Audi

I got the appeal of BMW and Mercedes when I was growing up. People would buy a BMW for something a bit edgy and sporty. They bought a Mercedes if they were older and a bit wealthier. The Audi demographic was harder to pinpoint. They seemed to be owned by alternative, quirky, slightly smug sorts. Smart but a bit . . . odd. You wondered if they all knew something that no one else did.

In 1979, Audi took a punt. They asked the FIA (the Fédération Internationale de l'Automobile, the world governing body of motorsport) if they would change the rules of the World Rally Championship to include four-wheel-drive cars. The FIA surprised everyone, including Audi, and said yes. Of course, by that time, Audi were already trialling a prototype with four-wheel drive and a turbocharger in a quarry in Nuremberg. That car became the Audi Quattro.

The Quattro debuted at the Geneva Motor Show in March 1980. No one could believe that something this cool was also a production car; it looked like a rally car. And I got to see it up close when I was still in single digits. Well, I got to see it whizz by anyway because in the early 1980s the Lombard RAC Rally went directly past my house. Everyone lined the street, kids sitting perilously on the kerb. The car we were all talking about was the Audi Quattro. It looked absolutely mega and it sounded incredible.

For three years running, between 1981 and 1983, the Quattro won the Lombard RAC Rally, with Finnish driver Hannu Mikkola behind the wheel for two of those wins. In October 1981, Michèle Mouton became the first female driver to win a world championship rally, in an Audi Quattro. Legendary German driver Walter Röhrl won the World Rally Championship in 1982 with Opel, but it was the Audi Quattro that won the manufacturers' championship that year. And in 1984. It was an unstoppable car. That's when everyone started to take notice of Audi.

The production Quattro models did have a weakness and that was the emphasis on straight-line performance. They weren't as agile as a BMW and not as good to drive, and part of the reason was the way the engine was positioned in the Quattro. Unlike other cars, where the engine was behind the front axle, in a Quattro, the engine was in front of the front axle, which meant they were nose-heavy and liable to understeer. Not that anyone cared – it just made you feel more like a rally driver!

Audi updated the Quattro for the 1990s, and in so doing they produced the first car to wear the 'S' badge, the S2 Coupé. It was stylish, aerodynamic and packed a punch with its turbocharged 2.2-litre engine, which they upgraded the following year, pushing it up to 227bhp. That model came with an overboost function – giving you a short-term surge from the turbocharger – to smash that overtake on a B-road. It must have been so satisfying in a street drag race, when your opponent is edging it and looking all smug, but then you deploy your secret weapon and fly by.

Audi, who had become a subsidiary of VW in 1966, eventually found their identity as a kind of posher Volkswagen. They carved out a niche in 1995 when they designed the A3. It didn't matter that it used the same platform as the Golf MK4, which came along in 1996. Audi had created the premium hatchback and neither BMW nor Mercedes had anything to rival it. The A3 also worked like a gateway drug for Audi. People started taking more notice of their performance cars and saloons.

Back in my previous life as a chartered accountant, one of my senior managers had an Audi A4 Cabriolet, with the silver windscreen surround. It was a beautiful car. His had a 3.2-litre V6 engine, which was quick, but they also did an RS4 Cabriolet, with a 4.2-litre V8 (the engine they used in the first-generation R8), giving you a ludicrous 414bhp. At that point in the 1990s, Audi were cranking through the gears, moving from niche to mainstream cool. And behind the scenes, they were developing something really cool: the TT. I remember the concept drawings for the TT at the 1995 Frankfurt Motor Show. It had such a pure shape, like a Bauhaus building had been reimagined as a car. But as with almost every single concept drawing of a car, there was no way they were actually going to make it. But Audi did. I think they changed the bumper and the rear window a bit. That was it.

Looking at the TT, you would have never guessed that it had the same so-so platform as the Golf MK4 and the A3. When the packaging looks that pretty, no one notices that the premium chocolate's exactly the same as the Sainsbury's Basics one. What people do notice, however, is if the chocolate hasn't been tested properly and causes your arse to suddenly lift into the air. At speed on German autobahns, Audi TTs were experiencing stability issues, which were intensified when the car's weight was concentrated on the front axle, as it is when you use your brakes. That's why the Audi TT ended up with a spoiler, so your car didn't turn into a Wright brothers' prototype and take flight. Some spoilers look incredible on cars. This one didn't, to the extent that people took the risk with the stability and chose not to have the retrofitted spoiler added. The desirability of the 'un-spoiled' version meant those cars ended up being worth quite a bit more, assuming those drivers *could* cash in and hadn't perished on an autobahn.

Aside from the stability issue, the TT wasn't a great car to drive, even in the 1.8-litre four cylinder turbo version with 222bhp, or a naturally aspirated 3.2-litre V6, which would give you 247bhp. But even the bigger engine wasn't as sporty as you hoped it would be. Part of the problem was that the Quattro all-wheel-drive system didn't fit the TT's transverse engine, so they used the electro-mechanical Haldex system. This meant that the TT was mainly front-wheel drive and only really started moving power around when the front wheels lost traction. The Quattro was mainly rear-wheel drive and supplied power to the front wheels when you needed it, which was far better for handling and stability.

The TT did get better and better with successive iterations, though. A pioneering feature of the second-generation TT was the digital driver's display. So you basically had your satnav as part of your instrument binnacle. No one else had done it before but everyone else copied it. Also, Audi led the way on exterior lighting and their daytime running lights design especially, which, again, everyone else nicked. By the time they retired the TT, twenty-five years after the original, it was a very good car. The TT transformed Audi's appeal. Also, it resonates with me because I became a motoring journalist the year the TT came out. So I've outlasted a design icon, although there may be an argument that the TT has aged better . . .

One of the things I really like about Audi is that they'll create a very good car, but then they won't be content with it. They want to keep pushing the envelope. The three-door liftback S2 coupé was terrific – sporty, cool, popular – but they wanted more, and they weren't afraid to ask for help. So they approached Porsche, who were struggling at the time (the early nineties) and welcomed the opportunity to build cars for other manu-facturers. The result was the first car in the RS series, the RS 2

Avant. It ended up as a marriage made in heaven, but the RS 2 Avant is more Cerberus than St Peter. It's an absolute monster. Even the logo's cool, with its four divided segments that make up a rhombus. On the left is the red, with the Audi rings; in the centre's the blue with the 'R'; and the 'S2' is in silver on the right. Underneath is another silver segment running along the length of the other three, with 'PORSCHE'. And inside you've got the Recaro bucket seats, anthracite napa leather and electric blue styling, if you chose it, which you should have done, to match that incredible blue exterior paint that everyone remembers. In a 1995 road test carried out by *Autocar*, the RS 2 Avant did 0–30mph in 1.5 seconds. That was quicker than the $815,000 McLaren F1. And we're not talking about a rival supercar. It's a bloody estate!

Audi's RS models got better and better. They run BMW M-series cars close and often are better than the equivalent Mercedes. I've had an RS 4, an RS 6 and an RS 7 and I've loved each one. They aren't one-trick ponies – they are good all around. For example, my RS 6, which was a 4.0-litre, twin-turbo V8 with 600 horses inside, would comfortably do thirty miles to the gallon on a cruise. They're super-efficient to the point that the accelerator will vibrate a bit to tell you to lay off the gas so you can coast down the hill.

My Audi RS 6 was such a fun, well-balanced, really big, very comfy and insanely quick car, but it had one big difference over a BMW M5. In an M5, you always know that you're driving a performance car. In an Audi RS 6, you forget. It's only when you really put your foot down that you remember that there's a beast under the bonnet. That's a good thing in some ways – maybe it appeals more to someone who likes keeping their powder dry until the right moment. But for me, it felt like I might as well just have the 2.0-litre diesel A6 and slap on the awesome

RS 6 body kit. I wasn't using the performance as much as I could in the RS 6, because I kept forgetting it was there. The car's not goading you to go faster, unlike in a BMW where everything about it says *Are you ready?* A BMW turns you into a racehorse at the starting gate.

There are some Audis where it's impossible to forget what you've got in front of you, like the R8. I had a long-term loan on a naturally aspirated V10 Spider, albeit with a green exterior and brown interior, so the depreciation must have been atrocious. But even in the dead of winter, that car was glorious with the roof down. The only downside of the R8 was that they had to move the bulkhead forward to fit the roof on, so if you're over 5 foot 11, you're not fitting in without having to fold your legs in two. But I'd have an R8 over a non-high-performance Porsche. It was better than the Porsche of the time – the 997 – and it moved the game on. I remember going to the R8 launch and speaking to the guy who was in charge of the project. I asked him what aspect of the car he would have spent more time on. 'The steering,' he said. 'It didn't quite have the feel of a Porsche.' There's never enough budget to develop a car so the engineers and designers are going to fully be happy with it. It doesn't matter how big or small the company, their legacy or pedigree, there comes a point where shit gets real and you just have to get that sodding car out of the door and off to the launch.

Audi had a genius former head of UK PR, and I'm convinced he was one of the reasons for Audi's huge growth rate in the first two decades of this century. He was instrumental in flying the Audi flag far and wide. It elevated him into exalted circles and bagged him an invite to the royal wedding (William and Kate) in 2011. The *Mail on Sunday* weren't happy about that, reminding readers that the royal motto is 'Dieu et mon droit' rather than 'Vorsprung durch Technik', but the royals do love an Audi. After

Princess Diana was famously papped driving around Chelsea in a turquoise Audi 80 Cabriolet in 1994, sales apparently doubled overnight. In fact, Audi seemed to be one of the few subjects Charles and Diana agreed on, seeing as Prince Charles had an A8 limousine and a couple of A4s for the servants. Sorry, staff.

William, Kate and Harry were all driving Audis in the 2010s (William in a 4.2-litre S4 and Kate and Harry both in A3s in case you're wondering). The rumoured up-to-60 per cent leasing arrangement discount reported in 2011 for the royal family couldn't have hurt, nor could Audi's sponsorship of royal charitable events like the annual Audi Polo Challenge. Audi's PR machine was firing on all cylinders. If a celebrity got out of a car at a major event, the chances were that the head of PR had negotiated it to be an Audi. He would even have a load of A8s on standby to chauffeur journalists, celebrities and royals about. At a time when Audi's product quality was increasing, he did a blinding job of increasing brand awareness and making Audi desirable. When done properly, PR is ingenious, cheap-ass marketing. You just need to get people who other people are looking at to use your product. The Audi PR guy was also forward-thinking and recognised the potential for alternative media channels. He was one of the first to offer us a long-term demo car at Carwow. We only had about 10,000 subscribers but he knew that this channel was going to be big.

Audi produced one of the cheapest and most effective TV adverts with their famous four rings commercial in 2009. There are four evenly spaced pegs on a wall and a hand puts four car keys attached to big silver key rings on each of the pegs. The first is an Alfa Romeo key with the subtitle 'Design?', then a Mercedes key with 'Comfort?', a Volvo key with 'Safety', then finally a BMW key with 'Sportiness'. Then the camera angles upwards slightly to focus on the four overlapping key rings together, which of course form the Audi logo. There isn't even any music. It's just a synth and a bloke banging what sounds like a yogurt pot. A marketing masterclass.

I've been on some crazy-ass trips with Audi. One of them was a press event in Mexico but there was some sort of legal hurdle at customs so the cars were all stuck at the border. I got a call from the Audi press office:

'I'm sorry, Mat. We've had to cancel.'

'Oh, OK, no worries.'

'But . . . we've already paid for everything – do you just want to come out to Mexico for a few days?'

I thought about the empty, free, five-star hotel room and what my dad would have said if I'd turned it down. So I (and journalists from other publications) headed out, but for compliance reasons we couldn't just have a holiday. We ended up working from there on other edits and also used the spare time with the Audi people to brainstorm and plan future video-shoot ideas – something we just wouldn't have had time to do had we been busy filming the car. And if you're going to be having meetings, it's nicer to do it in sunny Mexico than wet and cold London.

The pre-launch launch Mexico trip is what most people assume press events are like. The truth's slightly different. Here's the part where I'm going to find fault with my near-perfect job because, despite the fact that what I do is incredibly enjoyable and rewarding and there's no way I'd trade it for my old life as a chartered accountant – doing a proper job – I'm a motoring journalist and we're a whingey bunch. So, you get flown somewhere glamorous, which is amazing, but you're only there for about twelve hours. Your head's in a different time zone and your world's basically been turned upside down. Then you're up at the crack of dawn to be bussed away to the middle of nowhere with all the other journalists. These events always feel like an episode of *The*

Apprentice because you're competing with other media outlets for the right car, the right colour, the right spec and, most importantly, for as much time as you can get with the car. It's chaos. Then you try to get to grips with the car and shoot some half-decent content, surviving on the biscuits you've snaffled from the plane. Of course, your partner sees the Instagram story later that night when you're eating sushi and it looks like you've been pissing about all day!

Bentley

For as long as I can remember, Rolls-Royce concentrated squarely on luxury. Bentley went down a different road, and that road started on the race track.

Five years after they were founded, their first car, the Bentley 3 Litre, won the 1924 24 Hours of Le Mans. It was a beast – twice the size of the rival Bugattis that had basically won everything to that point. Ettore Bugatti, the carmaker's founder and big boss, called the Bentley 3 Litre the 'fastest lorry in the world'. Apparently, that was meant as a compliment. A Bentley 3 Litre won four successive 24 Hours of Le Mans races between 1927 and 1930. After the 1930 win, at the peak of their powers, Bentley gave up motor racing in what sounds a bit like 'Completed it, mate'.

The Bentley 4½ Litre followed the 3 Litre but it was the racing variant of the 4½ Litre that blew everyone away. That model was the Blower Bentley. Cubby Broccoli, producer of the James Bond films, might have stuck Sean Connery in an Aston Martin, but when we meet Bond in the books, he's driving a Blower Bentley. Remember the scene in the film *Casino Royale*, when Daniel Craig's Bond crashes the Aston Martin DBS V12 when trying to avoid Vesper Lynd, who's tied up and lying in the road? The same thing happens in the book, but the car is a Blower Bentley.

With the supercharger mounted ahead of the radiator grille, the Blower looks the business. In 1929 it must have seemed like something Captain Nemo dreamt up. It broke the lap record at

Brooklands, reaching 138mph, which must have felt like being fired out of a cannon back then. The track conditions weren't exactly the level concrete you might expect – in fact, the Blower was completely airborne several times. I've never driven a Blower, and that's probably a good thing given that they're worth about the same as a small country's annual GDP, but I did get my hands on one. I wanted to see how Bentleys of different generations stacked up against each other so I thought up a drag race with a difference. Stunningly, Bentley not only liked the idea, their press officer provided all the cars. And so we had ourselves the most expensive drag race in history (probably).

Representing Generation Alpha (born 2013–now) was the Continental GT Speed at £225k; firing up its engines for Generation Z (1997–2012) was the Continental R at £150k; the Turbo R, a snip at £48k, was the Millennials' (1981–1996) weapon of choice; and flying the flag for the Greatest Generation (1925–1945) was the 94-year-old Blower Bentley – yours for 20 million guineas. But, for its time, what a machine. My friend, drag-race partner and fellow car clown Yianni Charalambous and I squeezed into the back for a post-race race but the exhaust fumes were making us choke when we hit about 80mph, so we tapped out. We did, however, set a new world record: fastest contraction of lung cancer.

We do get spoiled by modern cars, in their silent, high-tech cockpits. But every once in a while, you want to smell the smoke and hear the roar. It's what makes being in a Spitfire more incredible than a Eurofighter Typhoon. The Blower has always been a petrolhead's wet dream so you'll have to imagine what the Bentley enthusiasts were up to when the company announced they were scanning all 630 of the Blower's components and creating a digital CAD model before bringing in artisans to build new models by hand. They even stuffed each seat with 10kg of horsehair, as per the original spec. The prototype alone took 40,000 hours of work. Bentley only made twelve new Blowers and each of them pre-sold for £1.5 million.

There are things that Bentley get spot on, like the way they feel to drive and the plushness of the interiors. The commitment to quality on the interior finish is matched only by Rolls-Royce. They used fourteen bull hides for the inside of each Bentayga. I also remember hearing that they make sure the fields housing the bulls destined to become Bentley interiors don't contain any barbed wire, so their hides don't have any scars.

When the Continental GT came out in 2003, it was the first Bentley made under VW's ownership. Like BMW launching the Rolls-Royce Phantom, the stakes could not have been higher. Cock it up and not only does your astronomical investment go up in smoke – you'd also be responsible for killing off one of the biggest car brands in the world. Throw in the fact that they're German custodians of a British cultural icon and failure would involve laying off a lot of British workers, and that's a recipe for pitchforks at dawn. Thankfully, though, the Germans know what they're doing. In 2003, BMW smashed it with the Rolls-Royce Phantom and so did VW with the Continental GT.

I went to the unveiling of the Continental GT at the Geneva Motor Show in 2003. I'd only been a motoring journalist for about eighteen months and I felt completely out of my depth because I was surrounded by people who could actually buy them. You know the feeling when you go to a party and you don't really know anyone there, so you end up constantly moving in circles so you don't look like a loser? That was me. So I left and jumped in a cab feeling a little bit like a fraud because I couldn't even afford to buy the badge on the bonnet, let alone the whole car.

As for the car, though, one benefit of wandering around in circles was that I did manage to get a very good look at it. It was an amazing-looking machine, the rear haunches especially. It just looked epic. It was a big leap compared to previous Bentleys in terms of the shape but it still had this Bentley feel to it – slick, solid, aggressive and managing to look fast without even moving. It's like Bentley perfected the next-stage evolutionary jump, and that's not easy to get right.

Underneath, though, the Continental GT was actually a VW Phaeton, the large, luxury saloon they produced from 2001 but probably shouldn't have because no one wanted a Mercedes S-Class from VW. They just bought a bloody S-Class. But all wasn't lost because the Phaeton's top-spec W12 engine (basically two VR6 engines) ended up underpinning the Continental GT and the Audi A8.

This was all part of Ferdinand Piëch's (industry legend, former engineer and chairman of the VW board) platform-sharing initiative and he had the bigger picture in mind. But it did mean that, dynamically speaking, the Continental GT wasn't all that. No one cared, though, apart from motoring journalists. It still had a big 6.0-litre twin-turbo engine producing 560bhp, four-wheel drive and, most importantly, the Bentley badge.

The Continental GT was the fastest production car in the world for a while, but I'm always sceptical of claims like that because when you actually read the article, you realise there are a number of 'qualifications'. Like it's the fastest four-seater production car, driven at a latitude of 60 degrees north, built on a Tuesday by a bloke called Bob.

At Carwow in early 2025, we bought a first-generation Continental GT (still with the VW Phaeton engine). It had 80,000 miles on the clock and we paid ten grand. When we looked at the car inside and out, we couldn't believe that you can pick up something that spectacular for that kind of money. There is a reason for that, though. If you do have a problem, like a sensor issue, you can't just run a software update or stick a new sensor on. The way the engine's been designed means that you have to remove the engine to fix it. So a £5 sensor ends up costing thousands.

The third-generation Bentley Continental GT (launched in 2018) shares its platform with a Porsche Panamera, which is a big step up from a Phaeton. It's so much more dynamic, so quick, so easy, so luxurious and so comfortable. It looks brilliant and it's a great car, but the complexity means that once it's out of warranty and hits the used market, the effect on your wallet is going to be seismic. That's why they depreciate quicker than the Soviet rouble in 1989. I thought Bentley did a really good job on the fourth-generation Continental GT hybrid in 2024. The V8 with the hybrid system is more powerful and quicker than the W12 engine on the previous model, although it doesn't give you the same deep and guttural roar of the W12 that makes you feel like you've just triggered an earthquake. But the fourth-gen is so fast and so impressive that I'm not sure I miss it.

In 2018, Bentley made a plug-in hybrid version of their Bentayga – their first SUV. Unlike Rolls-Royce, who treat the category 'SUV' like someone at Hogwarts has just yelled 'Voldemort', Bentley are prepared to admit that the luxury SUV they created for the luxury SUV market is in fact a luxury SUV. The Bentayga has got impressive road presence and they do sell well but, however lovely the inside, the shape and the performance, an SUV is still a bit of a workhorse. And with a car that's higher up, like an SUV, the manufacturer has to stiffen up the suspension because the centre of gravity is higher. It just means that they're never quite as comfortable as you think they're going to be. If I was going to go down the Bentley route, I wouldn't choose a Bentayga. It's a bit like a very expensive, delightful burger and fries. There's only so good you can make it. Give me the choice of any Bentley and I'd go for the Continental GT. That's Bentley at its best.

Bentley customers like to feel special and Bentley have perfected the art of capitalising on that with special edition models – something that sets you apart from the hoi polloi. And by hoi polloi, I mean paupers who can only afford a regular new Bentley. The Bacalar, based on a Bentley Continental GT convertible and created by their exclusive Mulliner division, features a redesigned aluminium and carbon-fibre body, all sorts of shapely new angles and lots of finery inside. But £2 million! And you can't get an airplane-size carry-on suitcase in the boot! It's partly about asset appreciation – cars in limited numbers do fare well in the future. But it's also about showing off. Sure, you could slum it in the VIP section at a Taylor Swift concert, or you could pay $5 million for a private event at your house. Super-rich people want something that no one else has. Like buying one of Van Gogh's eleven *Sunflowers* paintings. Or Twitter.

Talking of ludicrous extravagance, I've been on some crazy Bentley launches. One involved a road trip through Las Vegas and Death Valley, down through the Santa Monica mountains and into Malibu. The route's very well thought out, they put you up in nice hotels along the way and you're in a beautiful car. The idea is that they're creating a luxury 'experience' that a Bentley owner might want to have. The lifestyle and travel journalists who get invited along are in their element, winding through the Napa Valley and wafting on about sunsets and terroir. For them, the Bentley's a kind of upmarket hire car and any mention of it in the article is along the lines of 'my derrière was pleasantly cushioned by the voluptuous leather seating'. Meanwhile, the trimmings that the lifestyle journalists are literally lapping up are wasted on motoring journalists. As a car guy, I'd rather be driving or filming than eating foie gras and socialising. Although I may look back and think of the opportunities I've missed out on because I drove an extra 50 miles to be doubly sure that what I was saying about the ride quality of a Bentley compared to the equivalent Rolls-Royce was 100 per cent accurate.

I'm lucky – whenever I go on a launch, I'll take a camera person with me, but almost all the other journalists have to buddy up with another journalist in one car, and they take it in turns to drive. You're going to be in the car for a few days, so you really don't want to end up with someone you don't like or someone who's massively boring. But I'd happily take either of those types over someone who wants to show off how good a driver they are. Without fail, these people are terrible drivers.

I've heard tales of drivers being ordered to stop by their terrified fellow journalist passenger who gets picked up by the side of the road by a representative of the manufacturer. Sometimes, the bloggers and influencers who get invited on these trips are from parts of the world where a driving test involves driving fifty yards down the road and stopping. There was one such influencer on the convoy before my group on a Bentley launch and the line of Bentleys was being filmed travelling through this beautifully undulating landscape. The car at the back suddenly disappeared from view and was found on its roof with the wheels still spinning (the driver was fine). In order to avoid that sort of situation, a lot of the journalists try to find out who's going and pre-arrange who they're going to share with, so they don't end up with a 'last pick in the playground' situation.

Aside from playground antics and the occasional near-death experience, any Bentley event is special and everyone does have a lot of fun. Some folks have more fun than others, of course, and the Bentley 'Toy Box' event in 2020, which unleashed a bunch of journalists to play with their heritage fleet, was especially fun for me. My daughter was conceived during that event.

I blame Bentley.

I'd been filming abroad and flew back to the UK. I arrived at

Heathrow to find my partner Jo in a beautiful chauffeur-driven Bentley Mulsanne, which I'd be filming a review of the next day. I've got a soft spot for the Mulsanne. Everything's effortless – it's just got so much torque. They haven't got anything else like the Mulsanne in the Bentley range. It's pure Bentley. They even referred to it as 'The Grand Bentley' when they were developing it. They knew they were on to something special. Anyway, there was chilled champagne in the back of the Mulsanne and Jo and I were whisked off to the Cotswolds for a lovely meal in a luxury hotel. It felt like a honeymoon. And then throw in the fact that this took place during the final stages of the last Covid lockdown, when hospitality venues had just reopened and everyone was delighted to set foot in a shitty dive bar let alone a five-star hotel, and . . . what do you expect?!

It's testament to the luxurious, calming five-star Bentley experience that two people in their mid-40s, who were neither trying nor expecting to get pregnant, managed to do so. Bentley – the gift that keeps on giving. We came out of that trip thinking, *It was meant to be. We can take on a baby no problem!* And then you go to the hospital where they flatteringly label a pregnant woman over the age of thirty-five a 'geriatric mother to be' and you start to come back down to earth.

We didn't find out the sex of the baby until the birth and a few days later we decided on the name Grace. We were absolutely convinced it was going to be a boy during Jo's pregnancy. The name we'd chosen? Bentley.

BMW

Let's start with a good BMW gag.

What's the one part of a BMW that you never need to change?
The indicator bulbs: they've never been used.

I've had a lot of long-term-demo BMWs and there's one thing that's identical about all of them. Once I climb in, I park like a moron. I even started to think that their parking sensors were misaligned, which would explain why BMW drivers straddle multiple lines in a car park.

BMW created their M (short for 'Motorsport') division in 1976, but the first M car I really saw was the first 3 Series model, the E30 M3, when it was launched in 1986. It looked awesome. They've got one in the BMW press fleet, and it's cool inside and out – it even smells like the 1980s. The E30 M3 has shot up in value in recent years but they haven't aged well, mainly because it's a four-cylinder job, the handling's slow and the steering feels heavy, so it's underwhelming to drive. But the third generation model, the E46 M3, which came out in 2000, was a blinder. It's still one of the best cars around to drive.

Whenever a new design for a BMW comes out, I hate it, almost without fail. Like when you try olives as a kid. But then a few years pass, you give them another go and you get it. *Why haven't I been eating these?!* you ask yourself. BMW's designs have always been daring, innovative and sometimes controversial. When they made the jump from the E30 to the E36, I thought the E36 was horrendous. A couple of years later it became my favourite car. Then they launch the E46 and I'm thinking, *Why the hell have they done that?'* A couple of years on and I'm talking about it as one of the best-designed cars I can remember.

The limited-edition CSL (competition, sports, lightweight) version of the E46 M3, which was introduced in 2004, is one of my favourite ever cars. The chassis balance is perfect and it has exactly the right amount of power and size for the road. They'd also improved the steering and had retuned the SMG (sequential manual gearbox) so it was actually all right. When you're accelerating hard in the CSL, it gives you this little kick in the back each time you change gear, which is cool. The best thing about it is the carbon-fibre air box, which produces an incredible noise. I sometimes borrow the BMW press car and find some excuse to do some content on it, but it's mainly so I can give the car a spanking. Just writing this makes me want to borrow that car again. I just need to think of some content to do on it to justify the loan.

I remember that when the CSL was launched, it was only about ten grand more than the standard version and every journalist was saying that it wasn't worth the extra money. It's now worth three times more than the standard car, so we can all agree never to listen to motoring journalists when it comes to predicting investment opportunities. When you get into the standard model of the E46 M3 and then the CSL immediately afterwards, the scale of the difference makes you think witchcraft is involved. It was the same situation with the M5 CS over the M5 Competition. The CS may have only had an extra 18hp, but its lighter weight (70kg less) and upgraded suspension were game-changing. It was like a hare during mating season: aggressive, up on its toes and ready to sprint. I went on a track day with the M5 CS and a Porsche Cayman GT4. By the end of the day, I'd spent double the amount of time in the M5. It chewed the tyres to bits but it was more fun than the Porsche. That's a big deal.

I can't face buying an E46 M3 CSL for its current market value of around 110 grand, especially because I'm going to lose money as soon as I drive it, which I will (because that's what happens when you put miles on cars which are priced as investments rather than things to drive). A lot. I always think about the financial implications of anything I do with a car. Take the boy out of Walsall but you'll never take Walsall out of the boy. Also, I realise that worrying about the value dropping makes no sense, given that I almost never sell a car, but the knowledge that I'm pissing money away doesn't sit well with me. I think I'm pathologically loss-averse. So here's my plan to sidestep that financial landmine: get an E46 M3 and try to turn it into an CSL myself by fitting the steering rack and the air box and stripping out the interior. If nothing else, ruining a perfectly good car should make for entertaining content.

I went to the unveiling of the E46 M3 CSL in late 2003 but I wasn't senior enough at that point to review it. What I was allowed to do was to look at it. It didn't help that I was a boy from the West Midlands and I got the feeling that the higher-ups at *Auto Express* didn't trust me *in* cars, partly because it was quite a cliquey home counties vibe at the time and initially my face and voice didn't quite fit. I was like the keen kid in the classroom putting my hand up to volunteer for things, figuring that was the way to go. But it turns out that sometimes if you do ask, you don't get. To be fair on the management, I don't blame their hesitance. I learned that I wasn't there just to have fun driving cars but to create meaningful content for readers about cars they wanted to hear about. And that lesson has served me well as my career has progressed.

At *Auto Express* I was quite good at covering news stories and getting my pieces into national papers and on TV. I'd come up with ideas and we'd run features that would generate publicity for the magazine. As I discovered, the trouble with realising you're quite good at something is that it can be at odds with what you actually want to do. I wanted to drive the cars, but I was holding down the news desk. That's why I ended up shooting car reviews on video.

Video felt like a platform that was aligned with the way the world was travelling. Also, it suited me, because I'm a bit of a show-off. Plus, I could carve out a niche because none of the old-school motoring journalists wanted anything to do with something that didn't involve being shackled to a desk and scribbling like a tortured romantic poet. I wanted to inject a bit of humour and dynamism. I didn't like the old motoring journalism rules that everyone adhered to, like refusing to comment on the design of a car because 'it was purely subjective'. How a car handles is also subjective, isn't it? And what do you think is the main selling point of a car, how it drives or how it looks? If there's one thing that unites most car fans, it's the strength of BMW's design.

The life of a motoring journalist in the noughties was that of a rich poor man. You spent most of your time at an IKEA desk under strip lighting, propelled largely by Nescafé, sugar and fear of your editor, but the rest of the time you moonlight as a billionaire, being choppered into a luxurious location, eating incredible Michelin-starred food and spending the night at five-star hotels before being given the keys to a car worth hundreds of thousands of pounds. You fly home in business class (it's economy, though, nowadays, unless it's long-haul) and are very much brought back down to earth when you arrive at the airport because you ain't picked up by the trimmings train. You're standing in the rain wondering if you can afford the Heathrow Express, but who am I kidding – it's the Piccadilly line for me. You return to your dingy, one-bed flat in zone 3 and find your pay slip on the doormat. Then you play the game, *Did my meal and hotel last night cost more or less than my monthly salary?* There wasn't usually much in it.

When car manufacturers launch a car, they'll first hold the international launch, with the left-hand drive version of the car.

That's usually somewhere very pleasant in America, Germany or Spain, and the international motoring and lifestyle press get invited along. Later on down the line, the manufacturer holds their local market launch for the UK with the right-hand drive versions of the car, and that's usually held in the UK.

Sometimes the fancy foreign trip idea did backfire spectacularly for manufacturers if the car was shit, because, especially on a long road trip, the car's shortcomings are going to be exposed. This is part of the reason that influencers get pampered more than journalists, because journalists will be more objective and talk about the negative aspects of a product. Influencers (even if they're not being paid by a brand to advertise their product) tend to be overwhelmingly positive in their posts and reels. They're a better bet for a marketing department.

One of my first solo trips abroad as a motoring journalist was to test the second-generation BMW X5 over in the States. The first-generation X5 was a really good car. One of the reasons BMW bought Land Rover in 1994 was to utilise their four-wheel drive expertise and work up an SUV. They smashed it with the X5, which shared features with the Range Rover. I've got an X5 from 2002 at home and you'll find its V8 engine in the Range Rover from that period. The second-generation X5 looked a lot better and was a big step forward, to the extent that BMW didn't need to send us journalists on what must have been a very expensive trip. Although, I'm glad they did!

We arrived at BMW's assembly facility in Spartanburg, South Carolina. Once there, we found out the itinerary, which was a road trip down to Charleston on the Atlantic coast. Fantastic – lots of time with the car. After we'd set off, I found a thick envelope in the glove compartment. It was full of money. Being an ordinary lad from an ordinary part of the West Midlands and having a publisher for an employer, I spent the next two days back in my default setting of eating cheap-ass lunches and staying at dubious roadside motels (using my own money, which I was expecting to claim back through the publishing company). All the while, I wondered about the not insignificant sum of money in my pocket. Is this a test? Am I being bribed?

Then, on the final day, I called up the BMW PR department.

'Hi, yeah, um, what am I supposed do with this money in the glove compartment?'

[Long pause fuelled by disbelief] 'It's for expenses. You're supposed to use it for hotels and restaurants and stuff that you need for the trip. We just need receipts.'

'Oh.'

Ten minutes later, I pulled up at one of the grandest hotels in Charleston. A valet picked up my keys and I went inside to reception.

'Good afternoon, sir.'

'Good afternoon. Tell me – is this the most expensive hotel in Charleston?'

'I believe it's the second most expensive, sir. The most expensive is the Wentworth Mansion.'

'Hmmm. Could I have my keys back?'

I left the ridiculously expensive hotel and drove down to the even more ridiculously expensive Wentworth Mansion, where I tucked into a bucket of moules marinière. I stopped short of ordering a vodka martini and asking if Felix Leiter had arrived yet, but that was the vibe. I tried to justify this indulgence on the basis that if budgets aren't spent they are usually removed the next year. It turned out they would be removed anyway, as this kind of thing just doesn't happen any more. And I'm glad in a way, because even though I enjoyed the luxurious hotel, the cognitive dissonance of wanting to be impartial but then spending this money made me feel uneasy.

One of the problems you encounter when you're driving a car in isolation on a launch is that your opinion can be altered by, say, the quality of the roads you're on. If the roads are super smooth, you're not really going to get an idea of how the suspension performs. Also, if you drive a car and haven't tested the direct competitors, you might not get a sense of how it shapes up, in relative terms. That's what went on with the new M5 G90 hybrid, which I wasn't keen on first time round, but then something happened: I changed my mind.

The second time round, I tested the M5 G90 hybrid alongside the Audi RS 7 Performance and the Mercedes-AMG GT 4-door. There were still some things I didn't like about the BMW: it was big and heavy, but you didn't notice it compared to the others. It was complex, but it was versatile. It felt a bit more playful than the Audi or the Mercedes, plus, seeing as it's a hybrid, it's much cheaper on the tax front. All that's to say I was wrong. Sorry, BMW.

BMWs are generally best in class, but there have been a few missteps. There is no way I would buy a M5 G90 hybrid out of warranty, knowing what BMW's reliability can be like for second-hand cars. BMWs are generally quite complex machines and are built in such a way that when you need to replace something, you basically have to take them to bits. But with the hybrids, you've got the engine, the motor and all the fiddly bits joining the two of those up to worry about, not to mention all the ancillaries. There's too much to go wrong and when it does go wrong you're going to feel like you've just been picked up by one leg and shaken until anything of value falls from your person.

I had an i3, BMW's electric supermini that came out in 2013, and it was ahead of the game, really clever and looked so good. Most of the chassis and body was made of bespoke carbon-fibre reinforced plastic – the first mass-production car like it. The interior was incredible – it's all about tech and textures. The sloped wooden dash, made of some kind of sustainable eucalyptus, had such a beautiful shape and grain. And it didn't look like it was going to be really comfortable inside, but it was. Yes, the range wasn't great, but it came out more than twelve years ago – all of the batteries were shit then. It's an amazing, amazing car and I got to spend a lot of time in one on a road trip from London to Amsterdam. I had the range extender, which is basically a two-stroke motorbike engine that keeps up the charge level of the battery. I shot a video to see how the charging infrastructure shaped up along the way: good abroad, bollocks in Britain. The i3 was a terrific class-leading car, but then it felt like BMW squandered the advantage they'd made for themselves. Instead of the bespoke, innovative and pioneering approach they took with the i3, they started developing other models on the same platform as their ICE (internal combustion engine) cars. The results weren't as special.

While BMW's saloons and SUVs have always been top tier, their roadsters have been less so. The 'Bangle Era' was that period in the 1990s after BMW hired Chris Bangle as their head of design – the first time an American had been awarded that role. They produced all sorts of weird shapes and angles that I didn't like (at first anyway). But the Z3 was different. It was the first BMW designed while Bangle was in charge and I liked the look of it straight away. The front reminded me of a Messerschmitt cockpit. The problem was, and this is a deal-breaker when it comes to sports cars, it wasn't great to drive. For a company that was peddling 'The Ultimate Driving Machine' as a tagline, this one fell short. It felt like the car that people bought for the BMW badge. If you wanted a sports car that was great fun to drive, you picked the Mazda MX-5. Similarly, I'd never choose a Z3 over a Porsche Boxster. Sometimes BMW's manual gearboxes aren't the best – they're a bit springy and what you want in a sports car is a really tight gearbox, like you get in the Boxster and MX-5.

It took me some time to think of another BMW, other than their roadsters, that isn't that great, which is a measure of how consistently good BMWs are. I finally came up with the 2 Series Active Tourer, their small MPV, because it's a dull car for a BMW. It's built on the Mini platform, so it's front-wheel drive, and that feels like you're ripping the heart out of a BMW. And you can't avoid the question: *Why don't I just buy a bloody VW Golf?*

Some special-edition BMW models can be a bit hit or miss. The M4 CSL had a big price uplift over the standard model and it didn't feel massively better, but what do I know? It'll probably be worth three times as much in twenty years. To be fair, limited-edition models always tend to do well from an investment point of view seeing as they tick that ultra-exclusivity box. If it's a good car and it's scarce, it's a good indicator. It's just that sometimes we don't appreciate the incremental improvement a car man-ufacturer makes, in the same way that you don't notice those fifty press-ups you've been doing each day for six months until someone mentions that you look all right.

BMW have an absolutely massive fan base and they're probably the most viewed brand on the Carwow channel. BMW drivers do have a bad rep, but part of that is because their cars are a lot of fun to drive. For all their styling and performance finesse, what BMW offer when it comes down to it is really simple: they make cars for people who like cars.

Bugatti

Bugatti are confusing. When you hear 'Bugatti' you assume they're Italian. The cars look Italian. The founder, Ettore Bugatti, was Italian. But they're French. Well, they're based in France – in Molsheim, Alsace. Although, confusingly, when Ettore founded Bugatti in 1909, Molsheim was in Germany, but then two world wars happened and the town found itself on the French side of the border. One thing that stayed consistent was Bugatti's racing reputation. Their Type 35, produced between 1924 and 1930, notched up 1,000 race wins in that time. The only race Bugatti hadn't won was 24 Hours of Le Mans, but that would soon change with the arrival of the Type 57 in 1934, designed by Ettore's son, Jean. He was the prodigal son – a seriously smart engineer and a talented test driver. Jean's Type 57 won Le Mans in both 1937 and 1939. The 1939 car was driven by a French racing driver with a name that will be familiar to you: Pierre Veyron. Bugatti were on top of the world. Then everything fell apart.

On 11 August 1939, just two months after the Le Mans race, Jean Bugatti was testing a racing version of the Type 57. He was driving at about 125mph down a country road and had to swerve to avoid a cyclist. He lost control and ploughed into a tree. He died before the ambulance arrived. Nine months later, the German army invaded France and annexed Alsace soon after. Ettore was a smart cookie and had anticipated this, so he'd shifted as much of his machinery as he could to the relative safety of Bordeaux. The problem was, Hitler knew how useful Ettore Bugatti could be to his war effort. And that's because Bugatti's engineering genius knew no bounds. He'd created one of the world's first high-speed trains in 1933 and a pioneering racing aircraft in 1938 that might well have been the fastest plane of the 1930s. Planes, trains and automobiles. There was nothing Bugatti couldn't do.

The Nazis 'encouraged' him to start making vehicles for them and despite the 'we have ways of making you talk' undertone, Bugatti told them to sod off. The fact that he was vehemently anti-Nazi was part of the reason that he was forced to sell his company and all its machinery for under half its value, before upping sticks to his apartment in Paris with his family. After the war, he was basically accused of collaborating with the Nazis and his factory in Molsheim was seized by the French state. He sued but won on appeal, became a French citizen in June 1947 and had his factory returned to him the same year. He was eventually exonerated and found redemption at last. Unfortunately, though, he hadn't been in great health and that was his last fight. He died two months later. The business declined and stopped trading in 1952.

Italian entrepreneur Romano Artioli breathed life back into Bugatti in 1987 by acquiring the brand and building a factory in Modena. He brought the designer of the Lamborghini Miura (still regarded by many as the most beautiful car in history) on board to develop their first model. That car became the EB 110, which I remember well as a kid. Artioli named it the 'EB' after Bugatti's initials. A nice touch. That car had a sixteen-cylinder V12 engine with four turbochargers, a carbon-fibre body, scissor doors and it did 0–60 in 3.4 seconds. Plus, it looked so cool. They only made around 120 of them, one of which was famously owned by Michael Schumacher. I got to drive the one owned by the Petersen Automotive Museum in Los Angeles. They did warn me that this car always had technical problems, and sure enough it had some clutch and gearbox issues when we drag-raced it, so we only got one or two runs. It did feel pretty quick but, by today's standards, it wasn't earth shattering. The Ferrari F40 was much more of a thrill to drive. When the EB 110 came out, in September 1991, it was a tough market, seeing as the Diablo had been launched in early 1990 and demand for the F40 was still strong. Plus, much of Europe was in recession until the spring of 1993 and North America wasn't faring that much better, which was spectacularly bad news when you're making your first extremely expensive sports car and have thrown all your cash at it. Artioli liquidated the company just four years later.

One company not short of a few quid in the mid-1990s was VW. At the time, they were the biggest and most valuable car manufacturer in the world, under the stewardship of Ferdinand Piëch (see VW chapter), and they were at the top of their game. VW had been eyeing up the luxury market for a while. So they bought all of it. In 1998, they acquired Lamborghini, Bentley and Bugatti. VW were way ahead of the competition with their platform-sharing system, which was working better than any other car manufacturer's. This was their opportunity to showcase the best they could do as an organisation.

Under the new VW ownership, Bugatti made concept cars until they got to a point where they could put everything they'd learned together and start making a production vehicle. When it came out in 2005, the Veyron looked like a concept car. VW were flexing, big-time. And like the company's previous incarnation, they honoured Ettore Bugatti by adding his initials to the car's name, calling it the Veyron EB 16.4. Why the 16.4? It's a nod to the sixteen cylinders and four turbochargers on the Type 57 Bugatti that Pierre Veyron won the 24 Hours of Le Mans in back in 1939. These details mean a lot and it's something that VW respect. Once you've acquired a legendary car manufacturer, you've got a proud legacy to uphold. It's a tricky balancing act but if you can find a way to honour the past while designing the future, you're going to be on the right road.

It's some achievement that twenty-one years later, the Veyron still looks like a concept car both on the outside and the inside. It was completely groundbreaking. I remember walking to the office in central London in 2005 and seeing Simon Cowell's Bugatti Veyron. I stopped walking to stare at that car. As a junior motoring journalist, there was no hope of me getting into that car back then, but just being in the car's orbit was enough for me. The design was insane, the interior was magnificent and the performance was out of this world. The Veyron was the first production car to top 400km/h (249mph) and the world's fastest car until the SSC Ultimate Aero pipped it by 2.5mph in October 2007. Of course, three years later, the Veyron's big brother, the Veyron Super Sport, came along to destroy the Aero and take the record back. Pro racing driver Andy Wallace set that record. And the Veyron's first record. And the world speed record in the McLaren F1 back in 1992 for that matter. Andy's a legend. He's still Bugatti's official test driver, which is all the more impressive given that in 2028 he'll be a pensioner.

I've never driven a McLaren F1, but I know people who have, and they tell me that driving it at 220mph is pretty frightening. It's a completely different experience in a Veyron. And that's because the Veyron was built and tested pretty much in the same way that VW developed a Golf. It was conceived to be as solid and as easy to drive as a Golf, and it is.

I got to drive a Veyron for a drag race but it was having sensor problems, which meant it didn't launch properly. It was still epic. The Bugatti model that I have driven properly is the Chiron. The Chiron moved the game on again from the Veyron and you really notice the difference in performance because you're going from just under 1,000hp on the original Veyron to up to 1,600hp. I went on the launch of the Bugatti Chiron Super Sport. Well, I say 'launch' – it was more like a couple of journalists invited over to the factory in Molsheim. The site where they have their design centre and where they do their meeting and greeting with customers is a lovely place to be – it's chateau-in-the-country-with-deer-jumping-around vibes. The actual factory element of the site is off to the side and is super-modern.

At the factory, I got to review the production version of the Bugatti Chiron Super Sport, their top-speed model that did over 300 miles an hour. It's the fastest car I've ever driven on a public road. I did 212 miles an hour on the autobahn. I've done that sort of speed on the runway we use for Carwow videos, but a public road is a completely different proposition. For one, it's narrower and the visibility isn't so good, and then there's the small matter of other cars using the road. That makes everything seem so much quicker, especially on the two-lane sections of road. You have to look so much further ahead, to the limit of the horizon, basically. If there's a car in the fast lane at 80mph and you're approaching it at 200mph, you're essentially approaching a stationary object at 120mph.

The acceleration in the Chiron Super Sport is relentless and feels totally effortless from its 8.0-litre, sixteen-cylinder, four-turbocharger jet engine of a car. The way it pulls is absolutely insane, but the car feels totally stable. You trust it, so you feel comfortable travelling at ridiculous speeds. If you've ever had radio-controlled cars, you'll know that if you accelerate while you're holding the car up, the tyres balloon. The same thing happens on a car going very fast. At the Super Sport's 440 km/h (273mph) top speed, the tyres rotate over fifty times a second, subjecting them to forces of around 4,000g – the kind of G-force that could break a mechanical watch. Each of its tyre valves – which usually weigh around 18.3g – experiences centrifugal forces that effectively increase its weight to 55kg. There are so many dynamic considerations when you're creating a car like the Chiron Super Sport, but Bugatti nailed them and has made the car user-friendly and insanely comfortable.

I've driven more powerful cars that have been tuned but the car wasn't designed to travel that fast, so you are wondering, *Will the car stop? Will it explode?* Not so in a Bugatti. The only thing that made me ultimately back out at 212mph was the apprehension about what other road users were doing. *Top Gear* were on the exact same event that we were on, also testing the Chiron Super Sport, and their reviewer/presenter topped out at less than 212. Just knowing that I went faster was a nice, if slightly sad, personal victory.

Unfortunately, we also kind of annoyed Bugatti a little bit because I thought it would be a good idea to take the car to the McDonald's Drive-Thru. Not only is it one of the most expensive cars in the world, it's also more than two metres wide, which does present a challenge when you're rolling up to collect your Big Mac. Fortunately, in a spectacular underuse of someone's skills, Andy Wallace was on hand to help guide me through the Drive-Thru. When we got there, I deployed my exquisite French to ask the McDonald's chap how much the large fries were. Three euros! So I told him from the window of a Bugatti Chiron Super Sport that they were too expensive and I didn't have the money.

The biggest video that I've done is with the Bugatti Chiron racing a Red Bull F1 car driven by David Coulthard. I wasn't even driving. I was in the passenger seat, with the owner driving and me wildly gesticulating. David took us down on the quarter-mile but the mile was amazingly close. The way the Bugatti keeps pushing on after 100mph is insane. It's like it's riding a torque tidal wave.

Bugatti and Rimac became a joint venture in 2021 – Bugatti Rimac – owned by the Rimac Group (55 per cent) and Porsche (45 per cent). Mate Rimac, who watches and likes Carwow's videos, was given the reins at Bugatti, and in return, Porsche were allowed to up their share in the Rimac Group. One of the reasons this deal took place was because Bugatti were planning a replacement for the Chiron, and they were going to go electric. So who better to join forces with than Rimac, who'd designed the remarkable Nevera. As it turned out, despite the Nevera being an incredibly capable, extremely fast, beautiful car, Rimac was struggling to sell them. People don't want an electric sports car in the same way they want a sports car with an incredible internal combustion engine. It's a bit like the high-end watch market. Customers don't want a digital watch; they want one with gears, springs and dials. They want mechanical perfection.

Rimac understood all that and so Bugatti Rimac, for their second joint car (the first was the petrol-driven W12 Mistral, which was based on the Chiron but convertible, and was limited to a production run of ninety-nine), decided on a hybrid with an internal combustion engine as the centrepiece of the car but with the added performance of the electric motors. That allows the company to hit emissions targets and for customers not to piss off their neighbours with the roar of a Formula 1 car if they're just pootling about down the road. They named it the Tourbillon and I was one of the few journalists to go to its unveiling in June 2024.

The Tourbillon has a V16 engine and one of the problems with that is that it's longer than the more compact 'W' shape where the cylinders overlap. A naturally aspirated V16 is utter madness really, but one of the upsides is the sound it generates, and having heard the noise excerpts that have been released, it does sound a lot better than the Chiron. The Tourbillon is also heavier than the Chiron, though, because it's got batteries and electric motors, but they've done all sorts of trickery to save weight and prevent it from taking off. They've used computer modelling to make the parts and have 3D-printed some components. It feels like an evolutionary jump in terms of chassis because they've introduced weight-saving holes, a bit like you find in the structure of a bird's bones to make it super light. It means that it looks like a very cool alien exoskeleton. There are so many dynamic considerations but they've nailed them and made the car user-friendly and insanely comfortable. It's a car with a huge bandwidth.

Every other manufacturer seems to be focusing on high-tech screens inside, but Bugatti have gone the other way, employing a watchmaker to integrate visible metal cogs into the rev counter. It's such a great idea having that beautiful visual reference when the gear changes. The car's name – Tourbillon – is the part in a wristwatch that counteracts the effect of gravity, which in turn creates drag that affects a watch's movement. It's almost like, in an era of homogenised electric batteries and motors, Bugatti are going back to steampunk. They've embraced the power that electric motors generate but they've retained the bespoke, timeless craftsmanship unique to a high-end brand. It's a fusion of ultra-modern electric and mechanical excellence. And it sold out within hours despite costing £3.2 million. And that's just the basic spec.

BEST SPOILERS OF ALL TIME

1. **Porsche 992 GT3 RS** – This is an incredible wing that creates meaningful downforce. It looks so mad and moved the game forward on spoilers not just for Porsche but for all cars.

2. **McLaren P1** – The adjustable rear wing is just so cool, especially when it's in race mode. Also, it alters its angle to act as an air brake to slow the car down.

3. **Zenvo TSR-S** – Electronically activated, this wing can tilt depending on the corner you're going around, so that rather than just adding downforce, it acts like a wing, helping steer the rear of the car.

4. **Ford Sierra RS500 Cosworth** – This is a big, sporty wing that completes the look of the car. I'm not sure how effective it is in terms of downforce, but it is certainly effective in terms of making you want to buy the car!

5. **Ferrari F40** – Another wing that fits perfectly within the car's overall design and shape. It almost looks like a handle on a pram, but somehow it just works visually.

6. **Koenigsegg Jesko Attack** – The wing looks like it's been mounted on backwards, which is super cool, as is the way it's constantly moving to alter the aerodynamic performance and braking.

7. **Porsche 993 GT3 RS** – Swoopy and curly, it's a beautiful piece of design and even has vent holes to further optimise airflow.

8. **Aston Martin Vulcan** – The remarkable wing reminds me of the wings of the beautiful Avro Vulcan bomber. Absolutely insane.

9. **Bugatti Chiron Super Sport** – This wing lowers to reduce drag for high-speed runs and also acts as an air brake. Very cool.

10. **BMW 3.0 CSL 'Batmobile'** – The way this wing is sculpted to fit within the car's bodywork is chef's-kiss lovely.

Chinese Manufacturers

For all the Chinese cars I've driven, the only thing that they lag a little behind on (at the time of writing, anyway) is their driving dynamics. Recently I drove the BYD Sealion, which is a mid-sized SUV, and then the Tesla Model Y, back-to-back. In some ways, you can argue that the Sealion is better than the Model Y. I prefer the seats in the Sealion and some of the human–car interface. But then when you drive it, the Tesla feels notice-ably better. It was a similar story with another Chinese car – the Jaecoo 7, another mid-sized SUV, this one a plug-in hybrid. It's really good value for money and when you look at the car, it's got presence. The size and quality of the interior is good, as is the tech, but when you drive it, it feels like the kind of car that people who aren't into driving will buy. If price is the primary concern for the consumer, it's a good option and makes sense over the more expensive equivalent VW. But I need a car that drives well.

The way the car drives is now basically the only quality that differentiates Chinese cars from their European counterparts. I notice these things as a motoring journalist who spends a lot of time comparing similar cars back-to-back. If you just drive one of them in isolation, though, it's more difficult to assess. It's a bit like when you're choosing a sound system, and there's a good-value option that you literally like the sound of. You're almost ready to buy it, but then you try the one in the higher price bracket and it blows you away.

When the Chinese get it right, which they will, the European car manufacturers could be in big trouble. Even now, as it currently stands, the market is arguably unfair, because Chinese car manufacturers have much lower minimum wages, many of them are state-owned, which gives them significant financial advantages, and workers' rights are considerably weaker than in Europe. And yet Chinese companies are competing in the same market as European car manufacturers.

The Chinese have started making EVs from a clean slate, unlike European manufacturers, who often run two production lines, one for EVs and one for the ICE vehicles, which is costly. Sometimes, they use the same platform for both ICE and EVs, but this operation is a bit like a sofa bed: they're okay but you'd rather sit on a proper sofa or sleep in a proper bed. European carmakers are having to spread the cost of all these parts across lots of different models, so quality is being compromised in various areas. What the Chinese manufacturers are doing very well is producing good electric cars, broadening choice for the consumer and driving down the cost of electric cars, and that's welcome because they have been too expensive. Are they making cars for people like me? Maybe not. Are they cars that I'd recommend to people? Yes, they are and increasingly so.

Cars don't play as big part in people's lives as they used to. They're more functional. People used to get a lot more involved with their cars, partly because they'd break down a lot more, so even if you weren't into cars, you'd have to get better acquainted with your car. Cars now are more like just another white good in your kitchen, especially when they're electric. There's little to differentiate electric cars, but there are innovative ways of making electric cars more exciting – we've seen that with the Hyundai Ioniq 5 N's 'gear-change' system. As for electric hypercars, like the Rimac Nevera and the Lotus Evija, even though they blow most other cars away, they're not selling out. At the ultra-high end of the car market, people want the craftsmanship that's gone into an internal combustion engine. They also want the heritage of a well-respected brand. But as for normal people getting from A to B, if you're not that into cars, why would you not want an electric car? Putting the range anxiety and horrific depreciation aside for one second, they require less maintenance, are cheap to run, quiet, easy to drive, and even basic versions give you the acceleration of a hypercar. Anyone can now get that kind of performance and, for the best value, you're looking at a Chinese EV.

In China, they have their own version of YouTube, but we discovered that users were ripping our videos and uploading them. One day at a motor show, I had a Chinese guy come up to me and tell me that I was really big in China and that he'd like to manage me. Apparently, I have a nickname in China that is something along the lines of 'Cool little brother', although I think the direct translation is more like 'Donkey man'. We investigated and Carwow ended up doing a deal with a Chinese app for car fans that has become one of the biggest of its kind in China. This company took our videos, translated them into Mandarin and stopped all the other people who were ripping our videos. I remember going to a motor show in China where this Chinese app had a stand with videos of me playing. Loads of Chinese people came up to do selfies with me.

BYD (which stands for Build Your Dreams) is the world's largest manufacturer of plug-in hybrid and fully electric vehicles. They have now overtaken Tesla as number one for global electric car sales. And they've got big plans: to enable five-minute charging. For most people, the big thing that puts them off electric cars is range anxiety. If you can develop batteries that charge in five minutes, range anxiety will go out the window. This is where the Chinese can really step-change electric cars, and I have no doubt they will do. Ironically, it could be the Chinese that solve the problem with CO_2 emissions.

I'm not sure that in the UK and Europe we quite understand the scale that China are working on. When I flew into Wuhan, there were thousands of rows of massive apartment blocks. It was like nothing I've seen. There are cities in China that are built over loads of different levels, so you think you're at street level but then you look over the edge of the road and realise that there are another five levels below. It's like *Blade Runner*. When you see this kind of engineering marvel, you can well believe that five-minute charging is going to come out of China.

Citroën

Just like their unique suspension system from the 1950s, Citroën have had impressive ups and downs. They've produced two of the most iconic designs in car history, with the 2CV, which revolutionised rural France, and the beautiful, cutting-edge DS, which introduced never-before-seen features like directional headlights and disc brakes. Between these two models, Citroën flogged 10 million vehicles. And yet they filed for bankruptcy in 1974. How did it all go so wrong?

Part of the problem was that in the early 1960s, the workhorse 2CV and the DS executive sedan were selling at different ends of the market. They had no mid-sized, medium-priced car. Also, the French tax system punished vehicle engines over 2.0 litres with a hefty tax, so Citroën didn't have a decent, powerful, exportable engine. Things looked bright in the beginning of the 1970s with the success of the GS but they couldn't swallow the development costs for the ambitious range of cars they'd got planned. Throw in ownership wrangling and a major oil crisis and the company was sinking. And let's not forget the spectacular gamble with the unfortunately named Wankel helicopter engine. They developed the RE-2 helicopter prototype in just two years, but the engine wasn't up to scratch and started to overheat. Plus, the oil crisis had meant spiralling fuel costs. After Peugeot bought Citroën and formed the PSA Group in 1976, they actually pursued the helicopter project but it was never awarded government certification so the idea combusted. Their sole prototype was restored in the

1990s and I've sat in it in the Citroën museum in Paris. It's such a cool helicopter, and feels way ahead of its time, but it went the way of many Citroën concepts. Citroën remind me a bit of Marty McFly in *Back to the Future* telling the silent, stunned audience: 'I guess you guys aren't ready for that yet.'

After Peugeot came to the rescue, everything looked rosy for a bit with strong sales of the GS and the CX, but in the 1980s everything that had made Citroën unique was evaporating. They basically just produced worse versions of Peugeots. For the skilled Citroën designers, it was a bit like asking Le Corbusier to design a wheelie bin.

My uncle, who lived just round the corner from us in Walsall, had a white CX. He was a French teacher and loved everything French, but he drew the line at the beret, thankfully. I'd try to get round to his house after school but before he'd come home from work, just so I could see the CX pull in, stop and drop down on its suspension. It was the coolest thing. That was home entertainment in the early 1980s. Unfortunately, one day he thought he was in reverse but he was in drive – and he ploughed through the garage. While that car was still in one piece, my god the shape, the angles and the velour seats were incredible. It made the Fords and Vauxhalls the rest of my family owned look like baked bean cans – they were fine and they did the job, but that was it. The CX was a work of art, inside, outside and under the bonnet. One day, I'll buy one. I'd actually take a CX over a DS, which I also love, but I've driven one and they're a bit too old now and the brake is this bizarre circular rubber button, which reminds you of trying to get the tap to work in a god-awful train toilet.

In the same year (1970) that the GS, Citroën's four-door fastback family saloon, came out, they also unveiled the SM. This performance coupé was a different animal. Citroën had been working on it for ten years, under the codename 'Project S'. They were given a boost when Citroën bought Maserati in 1968, giving them access to some serious shit under the bonnet. The SM had all the styling finesse, aerodynamic mastery, tech wizardry and plush interior that Citroën had to offer, combined with an all-aluminium 2.7-litre Maserati V6 mounted behind the front axle. Unfortunately, though, the car managed to inherit the worst genetic traits of both Citroën and Maserati and became supremely unreliable. And as cool as the car looked, it does need to actually work.

Sales took a nosedive after 1971 and Citroën only produced 12,920 of the SM. Meanwhile, the GS was an instant hit and won European Car of the Year in 1971. The GS went through various iterations and sold more than 2 million in its sixteen-year production run. It didn't look as cool as the SM, but it got you from A to B.

The Citroën 2CV was developed in the 1930s and by mid-1939, they'd produced a pilot run of 250 of them. It had just been renamed the 2CV (meaning two chevaux/horsepower) and Citroën had printed their sales brochures in time for the Paris Motor Show in October 1939. Sadly, the Paris Motor Show ended up being slightly different than anticipated, consisting of hundreds of German Panzers rumbling down the Champs-Élysées. I did a TV show on the 2CV and learned from a chap in the Citroën museum in Paris that the prototypes were hidden in barns all over the country to prevent the Germans from finding them.

The 2CV is another car that I've completely U-turned over. Yes, the gearstick is ridiculous and feels like pulling on a bent umbrella. The silly fabric roof is utter nonsense. It's ludicrously slow and noisy. It's the kind of car a kids' entertainer should be driving. All that said, they're such fun to drive. They corner amazingly and perform surprisingly well across terrible terrain, which lends credence to the gag that the 2CV was designed so that farmers could travel across rough fields carrying a carton of eggs without any of them breaking.

I did an amazing trip with Citroën at a time in the early noughties when they were in the doldrums and producing some really dull cars. The Citroën press team were trying desperately to get coverage for their vehicles, however bland and un-Citroën-like they had become. At that point in time, the British Army almost exclusively used Land Rover Defenders for nipping around bases but, well, we all know what happens with Land Rovers. So they were on the lookout for an alternative option that would be cheaper to maintain and wouldn't die on you at a moment when *you could die.*

Citroën saw the opportunity and did a deal with the Ministry of Defence to replace the old Defenders with Berlingos, seeing as they were big, practical, and cheap to buy and maintain. The first I heard about it was when the Citroën press office called me when I was at *Auto Express* to ask if I was interested in doing a story about how the British Army were getting on with their new fleet of Citroën Berlingos.

'Sounds good. Where is it – Aldershot?'

'Mount Pleasant Air Base. On the Falkland Islands.'

What followed was an utterly insane trip. The Royal Engineers took us under their wing and we spent a few days hanging out with them. Of course, they treated us as new recruits, so we endured the full hazing experience. The official line from the upper echelons of British Army HQ was that they didn't conduct initiation ceremonies. But things were a little different on the ground. More liquid was consumed than at a horse trough in a heatwave, but of course, as you'd expect, that was the opening salvo. The main event was being locked in a jail cell, which only contained a heavy tank sprocket (think a hollowed-out version of one of the wheels that caterpillar tracks run along). I won my freedom after squeezing my body through the sprocket. I ended up looking like I'd been used as a sander.

One other lethal obstacle that you encounter in the Falklands is the drainage ditch running alongside the road to Port Stanley, the capital. It's ridiculously wide, deep and steep-sided and will swallow whatever vehicle you're travelling in if you catch it. The reason why such a ludicrously lethal ditch exists is because when the Royal Engineers dug it, they were working off statistics for rainfall that they thought were in inches. They were actually in millimetres. That's the story I was told, anyway. When you're with squaddies, you're never quite sure if you're being stitched up.

We took the Berlingos out, posed for some photos next to a Chinook, went for a walk, met some penguins and got insanely drunk. We flew back with a bunch of squaddies returning to the UK on a Boeing 747 that was at least fifty years old and had clearly been decommissioned multiple times. About ten minutes after take-off, we banked hard and after it levelled off, all the squaddies suddenly ran over to the right-hand-side of the plane. At first, it seemed like an emergency, but I saw what the fuss was about when I got out of my seat to have a look. What was actually happening was that an RAF Tornado had flown alongside us before turning upside down directly above the wingtip of our plane. Then it banked and did the same above the other wingtip. The pilot then gave us the bird and buggered off. So cool. It was a scene from *Top Gun* in real life. And I got to experience that thanks to a Citroën Berlingo.

My four-year-old daughter's favourite car is our Citroën Ami Buggy, which is such a fun, cool, little machine. Yes, it goes twenty-eight miles an hour but it's perfect for whizzing across my land, around the village and taking my daughter to school. It's the most eye-catching car I've ever had, over the Fiat 126 and even the 911 GT3 RS. It's also one of the rarest cars in the UK, seeing as there are only thirty or so around, which makes it rarer than my Porsche 911 S/T. It's Citroën at its best: quirky, innovative, desirable and completely bonkers.

Ferrari

I'm a Porsche guy, but you never forget your first love. And like all of my friends, it was all about Ferrari. It started the moment I saw Dean Martin and Sammy Davis Jr roll into shot in a Ferrari 308 GTS in *The Cannonball Run*. With its incredible low-slung sleekness in that Ferrari red, with the pop-up headlights, targa top, black slatted grilles and twin-round tail lights, it was the most beautiful car any of us had ever seen. It took some time to see a Ferrari in real life, though, growing up in Walsall, and it only happened once. That was a 328 at my local golf club. The inside was like a posh handbag, all beautiful tan leather, but you couldn't take your eyes off the prancing black horse in the canary yellow roundel. Hearing the sound of the engine, even when it was at idle, was like nothing else. That high-pitched roar felt like it was my new heartbeat.

I remember going to the British International Motor Show at the NEC in Birmingham in 1988, aged fourteen, with some friends. The only reason we went was to look at Ferraris, touch one, maybe even sit in one. The guys at the Ferrari stand were happy for us to take turns in the driver's seat. It had the strange ability to completely remove your powers of speech. All I could hear in my head was: *Oh my god! I can't believe I'm sitting in a Ferrari.* Teenage years were about obsessing over the next Ferrari to come out or appear on screen. That same summer, I saw *Ferris Bueller's Day Off* on video and the Ferrari 250 GT California Spyder (produced 1957–1963) from that film edged above the 308

GTS in my rankings. It's all curves, polished chrome and timeless sixties cool. You could see the whites of the eyes of roomfuls of teenage boys when that beautiful car smashed through that window. It's a replica, right? Tell me it's a replica.

Between the ages of fifteen and thirty-five, if you'd told me I could have had any car in the world, it would have been the Ferrari 250 GT California. And it would have been a smart financial move, because they're worth around £30 million now. To be honest, though, I don't even know if I would have driven it. I would have been too afraid of accidentally driving it through a window.

If I do one day own a Ferrari, it'll be in red. There's no debate about it. I'll agree with Enzo Ferrari on that. He famously said: 'Ask a child to draw a car, and certainly he will draw it red.' The *Rosso corsa* colour, or 'Ferrari red' as everyone who doesn't work for Ferrari calls it, is what a Ferrari should be. Along with those iconic Scuderia Ferrari yellow shields on the wings with the 'S' and 'F' lettering and the Italian flag at the top. These days, those shields are an optional extra and will set you back over £1,500. But if you've already spent north of £250,000 on a Ferrari, I guess it's loose change. I got a window into the mindset of a Ferrari customer when I spoke to the PR department at Ferrari UK many years ago about the options on a Ferrari I was going to drive. He stopped me short: 'It doesn't matter. It doesn't matter about the options. This is an irrelevance to our customers.' Ferrari owners don't care about spending an extra three grand for Apple CarPlay. Ferrari finally fitted Android Auto but the length of time it took them gave you the sense that they thought it was irritating; peasanty, perhaps. I think it only happened because someone who worked high up at Google bought a new Ferrari and then couldn't sync his phone. Ferrari aren't like other brands. They do things their way.

For example, it's been reported that so as to prevent 'flipping' for a quick profit, you can't sell your Ferrari in the first year unless you inform them first. You're also not allowed to make certain modifications to your car. Changing the prancing horse logo, adding a wide-body kit or changing the colour are big no-nos, as Justin Bieber found out. Officially he hasn't been banned from buying Ferraris again but he no longer has the right to purchase exclusive models and special editions. Sorry is not going to cut it.

Whenever Ferrari do a timed performance test against a competitor for any car magazine, website or YouTube channel, they want it to be the best possible example of that car, running in peak condition. So they'll insist on bringing over a team of engineers, who arrive much like an F1 team with tyres, gauges and all sorts of other computer equipment. The difference between Ferrari and Porsche in that sense is remarkable. When you tell Porsche that you'd like to drag-race their car, they just send you the car. Porsche also tend to downplay their figures on horsepower, 0–60 and top speed. It's like they build in a margin of error so your grandma could hit the numbers. With Ferrari, they want to control everything they can to make sure their car can do what they know it can do. They don't want to leave anything to chance. They want to win. And I think that comes all the way from Enzo.

For many years I had a good working relationship with the PR department at Ferrari. You sense that in the UK they don't have much autonomy and that the factory in Modena very much calls the shots. Enzo Ferrari was an unusual character, and that permeates through the entire company, a bit like Elon Musk and Tesla. Although, bear in mind that Enzo Ferrari has been dead for thirty-seven years, so the fact he's still having a big impact does say something about the shadow he casts. For decades,

Ferrari were a company where the 'generals' of the organisation were encouraged to compete with each other. It kept them on their toes, not knowing if they were going to get whacked by the big boss at any moment.

In the spirit of not doing things like other car brands, when Ferrari entered the SUV market in 2022, with the Purosangue, they refused to acknowledge that it was an SUV. Ferraris don't fit into categories. Unfortunately, I can't seem to stop calling the Purosangue the 'Poo sandwich', which isn't going to win me any mates in Modena. It's far from that, though – it's a brilliant car to drive. One of the key features is its naturally aspirated, high-revving 6.5-litre, V12 engine, which puts out a whopping 725bhp. Everything today, especially in a performance SUV, is turbo-charged, because you want the turbocharged torque, but the Purosangue's engine (the same engine as the Ferrari 812 Super-fast) differentiates it from its competitors, even the bespoke-platformed Aston Martin DBX. And it's got the theatre to go with it, like the chassis that drops down on its suspension in launch mode, the ridiculously fast-revving engine, the designer sculpted seats and the rear doors that open backwards.

At Carwow, we organised the ultimate SUV drag race in 2024, pitching a viewer's Purosangue against a Lamborghini Urus Per-formante, a Porsche Cayenne Turbo GT, an Aston Martin DBX707, a BMW X6 M, a Range Rover Sport SV, a Mercedes-AMG G 63 and a supercharged V8 Land Rover Defender. The Ferrari pipped the Aston to the post with 11.7 seconds for the standing quarter-mile.

After the drag-racing shenanigans, we did a brake test involving five or six brakes (with a long gap between each) from 100mph, and the brakes started squealing. The whole point of carbon-ceramic brakes is that you should be able to pound the crap out of them on a track. And we hadn't even done that.

Ferrari are going to reveal their first electric car on 9 October

2025. Every manufacturer's having to do one to bring their overall emissions down. Ferrari have built a vast new complex with the plan to make everything they need to produce electric cars in-house from 2026, but I wondered where they were getting their batteries and motors from before then. We knew that they'd used British manufacturer YASA to build one of the motors for their SF90 Stradale, Ferrari's first plug-in hybrid. We're talking a mid-mounted 4.0-litre, twin-turbo V8, plus three electric motors, 1,000hp, 800Nm (newton-metres) of torque and all in a package weighing 1,770kg. That's absolutely nuts.

I found out that Mate Rimac, who's a big petrolhead, owned an SF90 Stradale and I was keen (naturally) to drag-race one against a Rimac Nevera. He kindly supplied the Nevera. I wasn't expecting Mate to be the one driving it, but that's what happened. We shot the video at a disused airstrip in Croatia and, on the starting line, Mate told me the Nevera's stats: four electric motors, 120kWh battery, 2,360Nm of torque, 1,917bhp and the car weighed 2,150kg. He reeled off the info without thinking, but then again, he built it.

We did the drag race. The Nevera made the Ferrari SF90 (which even today is one of the very quickest cars we've ever drag-raced on Carwow) look like a Fiat 500. Twice. I asked Mate if I could try the Nevera. What followed was the sensation of being shot out of the barrel of a gun. Mate came over after I'd come back down to earth and saw that I'd managed 8.62 seconds for the standing quarter-mile. He was delighted and said: 'Then we have a world speed record for a road car. Congratulations, Mat. We'll have a beer tonight!'

Ferrari saw the video and were absolutely livid with me. Ferrari had actually stopped working with me before the drag race, but this was the final nail in the coffin for our relationship.

Fiat

I remember very clearly the first time I saw a Fiat 126. It was the late 1970s and I was outside my house with my parents. I'd never experienced nausea before when catching sight of a vehicle, but the Fiat 126 was the most repulsive thing I'd ever seen. And not just on the road. I remember thinking that if my mum had bought a Fiat 126 rather than a Mini, I'd refuse to go in it. Now, I have a Fiat 126 at my house, and I haven't bought it just so I can beat the crap out of it. I love it. What has happened to me and do I need to see a doctor?

A few years ago, I did a TV show about restoring cars, and one of them involved travelling to Poland to meet a guy who was doing up original Fiat 500s. He was working on a 126, which was basically the same as a 500 underneath. Poland made 33 million Fiat 126s between 1973 and 2000. You needed approval from the government just to buy one, which took about two years because the waiting list was so long. Of course, that meant that a lot of people, when they finally got their Fiat 126, sold them the next day at double the price!

He told me that in Poland you could pick up a good 126 for about £1,500, but a shit 500 was going to set you back at least five grand. Being a lad from Walsall and the son of a Scot, with a sharp sense of value for money, this piqued my attention.

'So I could drive a car that feels funny and funky, like a 500, but I just have to put up with a shit outfit?!'

He nodded.

'Sold,' I said.

That conversation transformed my life-long hatred of the Fiat 126. To be fair, value for money is a powerful motivator, and that's something that's never going to leave me. For example, I'll happily buy clothes from Asda's George range – they're cheap, quite comfy and I'll put up with the fact that they might look a bit shit. With George in mind, I was able to accept the design of the 126. Not only that – I started to like it. I began to feel that the design had aged really well, with its simple square shape.

I started dating a girl in Poland when we were out filming the TV show. She introduced me to one of her friends, who had a Fiat 126 and learned that I was a fan. He told me he could pick up a good one and arrange to have it done up for me. I doubled down and went for a beige one. The ultimate anti-cool car. I was planning to drive my new Fiat 126 back to England with a mate, which sounded like an awesome road trip, until it dawned on me that we'd be travelling at 50mph and it would take three days in the height of summer with no air-con and with the engine sounding like we had a food blender on full power, permanently. Plus, we'd be shoulder to shoulder to the extent that a fart leaving either of our bodies would vibrate through both the farter's body and the non-farter's body. I wasn't prepared to take that risk. Fortunately, my new Polish friend found someone who had a trailer who was prepared to transport it to my house for £200. That was the biggest bargain of the whole adventure. To be honest, I should probably sell it because I don't drive it that often and I'm a little bit worried that if I drove it in the wet, it would dissolve, seeing as the bodywork is basically made of light brown sugar.

The first thing you notice is how pure the driving experience is. It's the closest thing you can get to a go-kart. You feel everything. And it's so convenient in so many ways. You don't have to ask yourself *will I fit into that space?* because you will. It's also so small width-wise that you can fit through those speed bumps with the gap in the middle; no slowing down required. It still costs about £10 to fill up the petrol tank. Its turning circle is better than a London taxi. You could keep driving round and round a mini-roundabout if you wanted to give people any more reason to stop and stare at it, which they do all the time.

I noticed that a lot of lorry drivers were honking their horns at me. I asked one of them, who'd just pulled over, about it. He was Polish and told me that it's very common in Poland to honk as a mark of respect for their beloved Fiat, affectionately known as 'Maluch' (meaning 'little one' in Polish). That nickname was used so much in Poland that Fiat actually changed the model name to Maluch.

Yes, there are some problems with the 126. The fuel tank on mine reads back to front, which you never get used to; Christ it's loud; and lord help you if you crash because the crumple zone is basically your face. And there are the reliability issues, which did spawn some high-quality gags about the 126, two of my favourites being:

What's the heated rear window used for in a 126? As a winter hand-warmer for the guy pushing it.

What is the difference between the 126 and the 126 Sport? The Sport comes with a tennis ball on the tow hook.

I broke down in my Fiat 126 recently on a country road while running a quick errand. I had completely forgotten that I'd just been at a face-painting session with my three-year-old daughter. I'd painted her face and she'd painted mine, and I looked like a melted version of the Joker. I did wonder why it took me an alarmingly long time to flag someone down. Fortunately, a kind fella stopped and offered to give me a push. It was only when I got home and saw my daughter laughing at me that I realised.

The 126's predecessor, the Fiat 500, is a beautiful car, inside and out; the iconic city car. I've also got a soft spot for the Fiat Coupé 20-valve Turbo from the middle of the 1990s, which was quick in a straight line but shit around corners. It also generated loads of torque steer, so it was all over the place, but it was a really fun car to drive, with an interior designed by Pininfarina, the guys behind the Alfa Duetto Spider, the Peugeot 205 CTI and the Ferrari F50.

But the majority of the Fiat range was a case of *who cares?!*

In 2007, Fiat launched an updated 500 range. The high-performance model, tuned in-house by Fiat Chrysler subsidiary Abarth (Brits pronounce it 'A-baarth', but it is supposed to be pronounced 'Abbat', not that anyone's going to pay any attention to this explanation), came out in 2008. It's a cool-looking little car and there's even a limited-edition Ferrari version, the 695 Tributo Ferrari, and its 180bhp packs a punch. The trouble with the Abarth 500 models is that they're based on a simple platform – the 500 – and then they add rock-hard suspension and change the steering (which is nice and light on the regular 500) by adding artificial weight. Then they add the weight of a hefty price tag. Also, it's a disappointment to drive compared to competitors like the Ford Fiesta ST, which is superb fun.

I used to try to avoid attending Fiat launches in the noughties because they were a bit . . . odd. They would fly you out of Stansted to Bologna, then you were picked up by a coach to travel for several hours to the middle of nowhere. Once I finally got there, mid-afternoon, I got the impression that they were doing everything possible to avoid anyone actually driving the car so you didn't have as much time to notice their cars weren't great. It might all have been a clever strategy to tire you out so you didn't ask too many questions, go home and just give it an average mark.

Sometimes, I've found Fiats to be slow-burners, though. Take the Multipla, the look of which (much like the 126) used to make me throw up in my mouth. I used to think it was utterly hideous. Now, though, I wouldn't mind having one. Inside, they're cool with their three seats across the front. It's an innovative, different car, which has stood the test of time because people are still talking about it.

When I was working at Argos in the early 1990s and dreaming of owning a reasonably priced convertible (affordability has always been a criteria of my aspirations), the Fiat X1/9 was on my list. It was a wedge-shaped two-seater sports car, with a targa top and retractable headlights that looked similar to the Triumph TR7, although that Triumph didn't make the list. The three that did were the Triumph Spitfire, the Fiat X1/9 and the MG Midget. I liked the Fiat a lot because it was designed by Bertone, a really cool Italian design house. Also, it was mid-engined, which was cool because it gave the car the perfect weight distribution, which you usually see in an F1 car or a Ferrari – not an affordable Fiat. The trouble was, even at £750 or so, it wasn't affordable to someone getting paid £2 an hour at Argos with no luck scrounging off his parents, unless I worked at Argos for another ten years. Even if I could have bought one of the holy three convertibles on my list, I probably wouldn't have gone for the Fiat because a guy opposite us bought one. I never saw him drive the car, but I did see him spend a lot of time repairing it. Over time, it also began to corrode before my eyes, along with my dream of owning a decent convertible for under a grand.

BEST SMALL CARS OF ALL TIME

1. **Fiat 500 (1957)** – A car that got Italy moving. Designed for narrow Italian streets, it has a surprising amount of space inside. It has bags of character and it's one of the coolest-looking cars ever created.

2. **Mini Mk 1** – The original Mini was a ground-breaking car with its transverse engine and gearbox underneath. Incredible packaging and remarkable room inside for a car so small.

3. **Citroën 2CV** – Famous long-legged suspension designed to traverse a French farmer's field without shattering a dozen eggs on the passenger seat. I hated this car when I was younger, but since I became a motoring journalist, I grew to love it and now think it's one of the coolest, quirkiest cars ever made.

4. **Ford Fiesta (all ages)** – If you grew up in Britain, this was the small car that everyone had. Every generation of Fiesta has something going for it. It's a fabulous car and I can't believe they don't make it any more.

5. **BMW Mini Mk 1** – Had the cool quirkiness of the original Mini and was hugely fun to drive, feeling like a go-kart. The Mk 1 is the best of the bunch because after that the cars got bigger and were more *maxi* than *mini*.

6. **Renault Twingo Mk1** – We never got this officially in the UK but it's such a cool, well-packaged car. They love them in France and rightly so.

7. **Fiat 500 (2007)** – Continues with the cool-looking theme but with a modern chassis built on the Panda platform. It was a revelation for Fiat and turned their financial fortunes around.

8. **Renault 5 E-Tech** – If there's one car that will turn EV haters around it's this one. It's stylish, drives nicely and has got a quirky interior, if a little bit cramped. And for an EV, it's actually decent value for money.

9. **VW Polo (all ages)** – The German version of the Ford Fiesta, it's a great, practical, well-built car. Unlike Ford, VW haven't been foolish enough to cancel it.

10. **Suzuki Jimny (all ages)** – Its lightness and four-wheel drive system makes it brilliant off-road and great in the city because it's small and narrow but you're sitting up high so it's easy to park and manoeuvre. Feels a bit flimsy on faster roads when it's being blown about by the wind and consequently weaves around all over the place.

Ford

I think Ford hate me.

The most recent issue involved a video with a Transit van. Being the keen method actor I am, I spent weeks in the company of white van drivers to polish my performance. So I whacked on a hi-vis jacket, stuck the sunnies on and bish-bash-boshed a Ray Winstone impression. Laaarrvellly. I thought up a few consumer-focused tests for the Transit, VW Transporter and Peugeot Expert that would appeal to the van-driving demographic. First test: when frightening other vehicles at traffic lights or waking up rows of houses with a 6.30am delivery, which van sounds the most aggressive and rattly?

Ford thought that my implication that Transit van drivers were aggressive and inconsiderate was inaccurate. I'll just leave that hanging.

I would, however, like to give Transit vans credit for some of the finest comedy moments in British culture. Something special happens when a Transit hasn't been cleaned in a while. Their back panels transform into a canvas. And then the night artists come out to play.

'Also available in white.'

'I wish my wife was this dirty.'

The British motoring press have historically been quite favourable towards Ford, perhaps because Ford are a supporter of old-school, traditional print publications. They've never seemed to embrace Carwow videos in the way other manufacturers do – maybe they don't like our style. But I don't dislike Ford. When I was a kid, it was all about Fords. I loved my mum's Fiesta XR2i, my second car, after wrecking my dad's Mini Metro, which I appreciate sounds like a deliberate act to secure access to the better car.

In the mid-nineties, after I finished university and trained to be a chartered accountant, I finally had some cash, so I bought a motorbike – a Honda NC30 – and a white Ford Fiesta Mk2 XR2. I wasn't disappointed, which is a compliment because a lot of late eighties cars are crap to drive – the appeal for them tends to run on nostalgia fumes alone. I did enjoy the lift-off oversteer I used to experience going round roundabouts in Walsall. Back then, we were on the look-out for 'Cossies', Ford's collaborations with engine manufacturer Cosworth, which were pretty legendary in the mid- to late 1980s.

It started in the summer of 1986 with the Sierra RS Cossie (Cosworth) and its massive 'whale tail' spoiler. Who'd have thought that a humble hatchback could become the coolest car in town. Its 2.0-litre turbocharged engine generated 204bhp, it did 0–60 in 6.5 seconds and had a 149mph top speed. Just to outline how that stacked up, the latest Porsche at the time, the 911 3.2 Carrera, was giving you around 230bhp, a 150mph top speed and 0–60mph in 6.1 seconds (that 0–60 was according to the manufacturer but, in classic German style, that was probably a worst-case scenario; *Autocar* measured it at 5.4 seconds). You would have to part with £50k to get your hands on the 911 3.2 Carrera. The Cossie was £16k. It wasn't pie in the sky like the Porsche; that was an achievable aspiration.

But Cosworth were only starting to flex their muscles. Five hundred of the 5,500 Sierra RS Cosworths that were made were destined for greater things. They evolved into the RS500 and were fitted with larger turbochargers, intercoolers, airboxes and turbo inlet pipes, among other improvements, all designed to make them monsters on the track. The engine improvements boosted the bhp to around 220, the top speed to 150mph and 0–60 in 5.6 seconds. The RS500 was a brilliant car. Mental but brilliant. I would have loved to see it go up against the 911 3.2. What am I talking about – I can make that happen!

The Escort RS Cosworth, or 'Escort Cossie', came out in 1992 and it looked so cool, just like a rally car. It was actually a Sierra underneath – they just made the body bigger in certain places because the Escort's engine is a different shape and orientation. As a teenager, I knew a few people around Walsall who owned Escort Cossies and they weren't the sort of people you wanted to mess around with. I wasn't dodgy enough for an Escort Cossie.

I have had a number of farcical moments with Fords. The first took place in the driveway of our family home. I'd just bought a new Honda motorbike and I was paranoid about it getting nicked, so I moved my dad's 1988 Granada Scorpio in front so as to block it in. I didn't get the positioning quite right, which was annoying, so I opened the driver's door and started it up, but I was lazily sort of half in and half out, with one leg on the sill and the other on the brake. Only it wasn't the brake, it was the accelerator. I drove the car through our hedge and into our next-door neighbour's garden, heading straight for their pond. I managed to get into the seat and brake, but the front wheels had crept over the edge of the pond and it was balanced on its sills rocking backwards and forwards, a bit like the coach at the end of *The Italian Job*. Finally the tyres bit and I got back on to our driveway. After a few seconds asking myself what the hell had just happened, I knocked on my neighbour's door, apologised and started rebuilding the wall to the pond that I'd damaged. Fortunately, my dad was out with a mate so he missed all of this. He came back home and nothing looked out of place, because I'd got the scratches out with T-Cut. A victimless crime. I was a genius.

The next day, I got a call from my dad. 'Mat, what's happened to the footwell in my car?!' In my panic, I hadn't realised that the neighbour's pond wall had massively dented the underside of the car, creating what can only be described as a huge speed bump in the floor panelling.

He was especially upset because he loved that Ford Granada Scorpio V6. Everywhere he went in it, whatever the road and however long the distance, he seemed to drive it at the same speed: forty miles an hour. Driving into our driveway: forty miles an hour. M6: forty miles an hour. He had one speed. He was also notoriously tight with his money. The V6 version of his cherished Scorpio even came with air-conditioning, a fact that he was very proud of. We had a summer holiday in the south of France one year and went on the ferry and then drove down, mostly at forty miles an hour. Temperature-wise, we'd gone from the sort of standard, underwhelming sixteen-degree British summer, to at least double that by the time we'd headed past Lyon.

'Dad, can we have the air-conditioning on now please?'

'Absolutely not. We'll use more fuel if I put it on.'

'Pleeeaaassse.'

'The windows are open.'

'Pleeeeeaaaaassse.'

'We're on holiday. It's supposed to be hot.'

We spent the rest of the trip sweating buckets, all of which the super-warm velour seats soaked up. So we got back home and the next day, I went with my dad to the garage, because we'd been on a long journey and he wanted to check that everything was all right.

'Er, your air-conditioning's broken, mate.'

It turns out that the compressors had failed because it hadn't been used enough.

When I was a student, driving my mum's XR2i, the driver's

My father always liked posing with his cars. He was born a little too early for Instagram, but here's a picture of him with his first car, a Singer Gazelle Series 3 (*right*), and one of his favourites, a Ford Granada Mk2 2.3L (*below right*).

At the time of writing I own thirteen cars. Some are in the picture (*above*) – the Porsche 911 996, Suzuki Jimny, Citroën Ami Buggy – but I've since acquired a GR Yaris Mk2, and the 911 GT3 RS has been replaced by a 911 S/T.

My top 10 game-changer cars are: Ford Model T, Tesla Model S, VW Beetle, Mini (Mk1), Porsche 911 (*bottom*: me with a G-Body), Toyota Prius, VW Golf GTI, Audi Quattro (*top*: Group B Rally Car), Ford Mustang, Nissan Qashqai.

My best hot hatches are: Toyota GR Yaris Mk2 (*top*), Honda Civic Type R EP3, Renault Sport Mégane R26.R, VW Golf R Mk8, Renault 5 GT Turbo, Honda Civic Type R FK8 (*bottom*, along with a VW Golf GTI MK8 Clubsport), Peugeot 205 GTI 1.9, Ford Fiesta ST Mk7, Ford Focus RS Mk1, Mercedes-AMG A45 S.

My top 10 spoilers are: Porsche 992 GT3 RS, McLaren P1, Zenvo TSR-S (*bottom*), Ford Sierra RS500 Cosworth, Ferrari F40 (*top*), Koenigsegg Jesko Attack, Porsche 993 GT3 RS, Aston Martin Vulcan, Bugatti Chiron Super Sport, BMW 3.0 CSL 'Batmobile'.

Me with an Audi RS6 GT estate. My other best fast estate cars include: BMW M5 Touring (E60), Mercedes E 63 Estate (S212), VW Golf R Estate, Subaru WRX STI Wagon, BMW M3 CS Touring (G81), Volvo 850 T5-R, Porsche Taycan Sport Turismo, Citroën CX Turbo Estate, Skoda Octavia vRS Estate.

Volvo used to be the king of the estates. They now only sell SUVs and EV saloons in the UK, though they are as safe as everyone says they are. According to the UK government's crash records, which have been going since 2004, as of May 2025 not a single person has died in the UK in a car-to-car accident in an XC90.

My top 10 hypercars are: Porsche Carrera GT, Koenigsegg CC850 (*top*), McLaren F1, Gordon Murray T.50, Bugatti Chiron, Pagani Zonda R (*Pagani Huayra R Evo pictured bottom*), McMurtry Spéirling, Ferrari F40 (also in top 10 spoilers), Mercedes-AMG ONE, Aston Martin Valkyrie.

The McMurtry's unique fan-powered downforce system generates two tonnes of downforce – even when stationary – which basically sucks the car to the ground, even if the car is upside down.

My first time driving a Bugatti Chiron Super Sport was taking it on the autobahn to go as fast as I could with their legendary development driver, Andy Wallace. And then going to a drive-through McDonald's!

The iconic 1992 McLaren F1 car broke numerous speed records when it was first released, topping out at 240.1 mph.

The Valkyrie is like a Formula 1 car – it's all aerodynamics and downforce wrapped up in stunning packaging. And that makes sense – it was designed by engineering legend Adrian Newey, who's picked up twelve constructors' championships in F1 and is now the technical director of the Aston Martin team.

door lock suddenly stopped working. The only way to get the door to shut was to sit in the driver's seat, open the window and lock the door from the outside. One day I was reversing out of the drive and forgot to go through the whole window-opening, reach-around fun so the door swung open just as I was going past the tree on the edge of the driveway. The tree took the door clean off. It did solve the locking problem, though.

Then there was the white Mk2 XR2 I bought after uni when I found myself with some money while working for PwC in the City. The oil warning light on that car had been on for ages, only I didn't know it was the oil light – it was just an orange light, and I was in my early twenties, so I didn't think that much about it. The car still worked fine. And then one day when I was driving home from work, the car started making a really bad noise and conked out. I rang the AA and one of their guys found me.

'You've . . . you've got no oil in it, mate. Your engine's dead. I'll stick some in and see if we can get it going but I wouldn't hold your breath.'

'Shit.'

He put some oil in and it started it up all right. I had it for another two years with no problems at all.

I started as a motoring journalist not long after Ford had brought out the Focus, and that car was a game-changer. It had this clever suspension system that made the VW Golf feel like a horse and cart. The Focus was so good to drive, and it was the same with the Mondeo. Both of them were absolutely brilliant cars. And then they produced high-performance versions, starting with the Mondeo ST24, with its upgraded engine, lower road stance and sports seats which were all well received. But it was the Ford Focus RS Mk 2, with its 305hp, 2.5-litre, five-cylinder, turbocharged engine that pushed the game on. That car was wild.

The Focus looked really modern when it came out and the car did exactly what you wanted it to. You felt really in control of it, almost at one with it. And that was all down to the way it was engineered. Everyone knew when they drove a Focus that it felt special. It made you feel alive and there was nothing to touch it on the market in that category. Ford had their huge marketing machine so people were taking interest, the look of the car was pulling people in and then when you test-drove it you were sold. People who had one were raving about them, and so their mate would buy one. Put all that together and it's no wonder that Ford were dominating the market.

I went on the Mk 2 Ford Focus RS launch and that involved a road trip in the south of France, along the 'Route Napoléon', the 100-day route he took through France on his return from exile on the Mediterranean island of Elba. The car was brilliant, really fast and looked amazing. But this was a good example of the inflated impression you can experience driving a car in isolation along beautifully smooth roads. I found that out several months later when I tested two cars together in the UK: the Focus RS and the high-performance Renault Mégane R26.R.

The Ford had the bigger engine, more cylinders and more power. It looked like a rally car with its massive wing. The Mégane R26.R looked more like a touring car, with the big spoiler, titanium exhaust, crimson wheels, and stickers and stripes along the bodywork. I was convinced it would be an easy victory for the Ford, but I was shocked to find that dynamically the lighter, more responsive Renault shat on it. It was quite some transformation when you consider that, as a base car, the Focus was the better starting point than the Mégane. I got in the Ford and it felt nose-heavy, a bit overweight and a bit soft compared to this precision weapon of a car. The Ford reminded me of the big guy at the pub who everyone's scared of. And he starts picking a fight with this smaller, wiry guy, with his cool sporty coat and shoes. Of course, the smaller guy stuns everyone by giving the big guy a walloping because it turns out he's in the SAS. You never know what's under the bonnet.

Fast-forward to now and Ford have only got one cool, fast version of one of their cars – the Puma ST – unless you include the Mustang, which is a great car but it's very, very niche (in the UK and Europe anyway). The price of a Mustang in the US is a bargain compared to what you pay in taxes and tariffs when you import them to the UK. It means that the money you'd spend on a Mustang in the UK in 2025 is the same as a Porsche Cayman. I know which one I'd go for. I do like the Mustang and the V8 engine is great – I've just never had the religious experience that some people do with Mustangs. But when you're in the States, you can't resist a convertible V8 Mustang as a hire car. It's mandatory. Yeah, it's big and not practical but you can't argue with the road presence and sound. It is a bit of a one-trick pony, but what's wrong with that for an occasional outing.

It's the same with the Ford Capri – Britain's answer to the original Ford Mustang, and there's a lot of love for it, especially the 280 Brooklands with its 2.8-litre, fuel-injected V6 engine and limited slip differential. My mum's mate had one and I remember the bucket seats and check patterned interior. It was really cool. The new 2025 Capri, though. Why? What a waste of that badge. A lot of people are too young to remember what a Capri was, in which case, what's the point in calling it a Capri? There are so many names they could go for instead. I could think of a few. The only reason to use 'Capri' is for the nostalgic vibes, but the new one is a small, underwhelming SUV EV, with no connection to the original Capri whatsoever except for the name. It's like the remake of *The Italian Job*.

Writing this has reminded me that I should review it at some point, though, if only so that I've reviewed and driven every car on sale, because, well, that's part of my job!

Honda

Honda and Toyota have been really good at different times in their history. It feels like one's always edging ahead of the other and that means that my favourite keeps changing. But one thing that's stayed consistent about Honda, for me, is the quality of their engines. Honda engines are very clever, very complex and very reliable. Back in the day, my mate's mum had a Honda Civic with a flat back and a 1.6 VTEC engine. Normally, when you drive a car, you put your foot down, the revs rise and the car gets faster. Hondas rev quicker and then, suddenly, towards the top of their range, they pick up even more. That feeling reminds me of riding a Honda NSR125 two-stroke motorbike, where you'd twist the throttle and feel nothing, nothing, nothing . . . and then they hit that power band and the engine comes alive. Also, I love that some Hondas masquerade as granny cars, but there's something special inside. They're a bit like the Clark Kent of the car world. It looks like an average, boring little car. But if you know, *you know*.

Honda smashed it in the early 1990s. The first one that caught everyone's eye was the wedge-shaped second-generation CR-X VTEC. It was absolutely brilliant. I loved the look of it, the interior design, the seats, and that special VTEC engine. They're really expensive now, but that is one car I would definitely love to buy. I think I'd take one over a Mk1 Toyota MR2, because even though the CR-X is front-wheel drive, it's all about that 150bhp VTEC engine. It's funny when you do drive these cars from the 1990s because they're not as quick as you remember. Maybe because your late teens and early twenties are such a seminal part of your life, everything seemed more colourful, faster, more vivid. Well, it certainly did when I was riding around on a Honda NC30 VFR400 motorbike at ridiculous speeds. That thing would rev to 15,000rpm. It was a V4 engine with four gear-driven camshafts rather than chains. It had a single-sided swing arm, so it looked like a Ducati, but Honda did it long before Ducati! My mates were into Kawasaki and Suzuki, and the bikes were pretty much the same money as the Honda, but it felt like twice the amount of development had gone into the Honda. Honda had more money to develop it and so they could make it more comfortable, reliable and more exhilarating to ride. It looked great, sounded great and went great. The Kawasaki might have edged the Honda in one or two areas, but the Honda was the complete package. It was the Porsche of the bike world.

It was the same with the Honda NSX. You feel like you're in a fighter jet, and you can't get much cooler than that. It turns out that's exactly the feeling they were going for: chief designer Masahito Nakano had based it on the bubble canopy of an F-16 Fighting Falcon. They even brought in Ayrton Senna, whose Formula 1 McLaren was powered by a Honda engine, to help with the development stages. He tested the NSX at the Nürburgring and other F1 circuits, including Suzuka in Japan, and his feedback stiffened the chassis and made adjustments to the suspension and handling. There's a terrific video (seriously – look it up) of Senna testing the NSX at Suzuka wearing sunglasses and a smart zip-up jacket, but the highlight is the pedal cam – and not only so you can see his inputs while he's gauging the limits of grip. It's because he's wearing high white socks and a pair of tan loafers.

When the Honda NSX rolled out in 1990, it blew away the Ferrari 348, which had come out the year before. The Honda was even better looking and could absolutely destroy it on the track.

And at the heart of the NSX was that amazing-sounding 3.0-litre V6 VTEC engine. It was a stunning car.

The Civic Type R EK9 came out in 1997, with its red background on the 'H' badge identifying the beast beneath. Gone were the days of the Civic looking like a granny car – this one came with blood red Recaro bucket seats and titanium gearstick. But it was all about that VTEC engine. Unfortunately, though, they only produced the EK9 for the Japanese market. Occasionally you'd see an import model, and when you did, you and your mates would be asking the same question: when the hell are they going to bring out a UK version?

Three and a half years later was the answer, with the arrival of the EP3, made in Swindon. It was affectionately nicknamed 'the bread van' for looking like a loaf of Warburtons, and although it didn't have some of the features everyone loved about the EK9, like the red Recaros and helical limited slip differential (LSD), this was no granny car. Not with its 2.0-litre VTEC engine and massive exhaust. They ended up putting back those fan-favourite Recaros and the LSD (along with upgrading the engine, chassis and exhaust manifold among other things) for a Japanese domestic market version of the EP3, only they didn't make it in Japan. They made it in Swindon, and exported it to Japan. It was an absolutely brilliant car. And it made history: finally something good came out of Swindon.

Then came the third-generation model of the Civic Type R – the FN2 – which used the same engine but had a freaky-ass design, which some people didn't like. I wasn't one of them, though – I thought it was incredible. The next generation, the FK2, was the first Civic with a turbocharged engine, and my god that was fast. I've liked every single version of the Type R Civic. The FN2 was the most 'Wow' visually but the least 'Wow' mechanically, because it wasn't actually as good as the EP3. The third, fourth and fifth generations were all special too and I've had each of them. Since 2007, the only Type R Civic that hasn't been on my driveway is the latest version, the FL5, which came out in 2022, but I've driven it. And guess what? It's an amazing car. The Honda Civic Type R is like one of those rare albums – every song is great.

I remember filming some drag races with Civic Type Rs we'd got from their UK heritage fleet, and there was a weird dynamic going on between the press office and the people who ran the heritage fleet. The press office just want to get coverage for their assets. But the people who look after the heritage fleet treat each car like they're one of their babies, so they're very cautious and protective. So when the press office made the heritage fleet team hand the cars over to us to drag race, I could see them wincing as we were launching them. After we'd done the drag race, one member of my team took the car around the track and he really flung it into the final corner, trying to make the car go sideways. He succeeded, but it went sideways on to the short grass, on to the longer grass and then into a big patch of mud. It was like someone had power-washed it with slurry. You should have seen the look on the Honda heritage guys' faces.

After the Honda heyday in the 1990s, just like Toyota, Honda started to make dull cars. They moved away from performance (or fun) models and it was all CR-Vs, Jazzes and normal Civics. Dull, dull, dull. And with import duties and the yen being strong, these dull cars were all quite expensive for what you were getting. So it became a case of *Why would I have one of these cars?* There were some exceptions to this boring procession, like the sporty Honda S660, which came out in April 2015. I'd always liked the S660 – it fitted into that cool/unusual category I like, plus it reminded me of a baby NSX.

In late 2024, at a car auction in Japan, I bought an S660. I'm going to tune it, because it's a little bit slow with its 66bhp, but you can get them quite easily up to 100 and then they're really quite quick. It's a tiny, light car. The brakes feel so sharp and the steering is really responsive because it's such a light, agile little creature. And then there's the location of the engine, just behind you, so it's like a racing car, and you notice that especially when you chuck it into a bend, because it's so manoeuvrable. When you go into a corner in a car like the S660, it's not a case of turning the steering wheel, which causes the rear of the car to follow the front wheels. The whole car is rotating slightly around its front wheel, and that means that you don't need to turn the steering wheel as much to get the car turning. It's like you and the car are exactly in sync. You've found the sweet spot, and this is at the heart of what makes certain cars more fun to drive than others. It's why, for a middling driver like me, a Porsche Boxster is a better-handling car than a 911. The Boxster's weight distribution is perfect – the engine's just in the right place and it feels like you're dancing around a corner. In a 911, the engine is in the back, so it's a bit harder to get the car rotating, and when it does, it can go too much. As for the Honda S660, even though it's slow and small, it does that thing that all good cars do: it does exactly what you want it to and it tells you what it's doing. Another thing that sets the S660 apart is just how unusual it looks. I get more attention driving a S660 than I do driving a £250k Porsche 911.

At a car auction in the UK, the cars are driven in one at a time and people bid either at the site or electronically. In Japan, you're in a big hall surrounded by digital screens. No cars actually appear in front of you. Instead, on these big screens, there are about twenty cars. And that's because twenty auctions are going on at the same time. You sit in front of a couple of smaller screens and get to hold this joystick. It's like a computer game and, like a computer game, everything happens very quickly. Your ability to take in all the information and the speed with which you bid can determine the price that you pay and whether you pay too much or too little. It's nuts.

The upshot of all this high-stakes excitement is that I'm addicted to buying cars in auctions in Japan. There's content and profit to be made! Generally, at Japanese auctions, when buying a car more than ten years old, you do feel like you're buying with less risk than you would do in the UK. First off, the way the Japanese care for and maintain their cars is something else. People actually buy second-hand European cars and import them from Japan because they've been looked after so well. This kind of attitude extends into the way the auction house treat the cars they sell: they hire expert engineers to go around and test the cars and give them each a rating. While you can never be 100 per cent sure, it does give you a pretty strong idea of the quality of the car. I do wonder what the Japanese must think of our auctions, though. At the bottom end of the market, you really can buy any old cack at a UK car auction. It must all seem like a Monty Python sketch.

BEST HOT HATCHES OF ALL TIME

1. **Toyota GR Yaris Mk2** – A rally car homologation special for the road with a completely bespoke body. It has a clever four-wheel drive system, a crazy 1.6-litre engine and it's basically like you're driving double WRC champion Kalle Rovanperä's rally car.

2. **Honda Civic Type R EP3** – Affectionately known as the 'Breadvan' because it was a bit of a granny car, but the mad, 2.0-litre naturally aspirated engine with VTEC (Variable Valve Timing and Lift Electronic Control) gives you an incredible kick up in power when you go past 6,000rpm.

3. **Renault Sport Mégane R26.R** – Basically a touring car for the road. It doesn't have rear seats, but it does have four-point harnesses, which are a pain when you're driving it normally but, oh my god, this car grips the road like nothing else.

4. **VW Golf R Mk8** – Everyone goes on about the Golf GTI but the Golf R is the better car. The latest version, with its clever four-wheel drive and rear differential, which can send power to the outer wheel with the most grip, allows you to drift this car nicely. So much fun, so capable and does everything well.

5. **Renault 5 GT Turbo** – Out of all the eighties hot hatches, this is my favourite. It just looked right and had plenty of performance back in the day with its 1.4-litre turbocharged engine.

6. **Honda Civic Type R FK8** – The second turbocharged version of the Type R but this one has an amazing chassis and a brilliant front end with its limited slip differential. It goes around corners like it's on rails.

7. **Peugeot 205 GTI 1.9** – Many people say that this is the best hot hatch of all time. It's a car that likes to do lift-off oversteer, so if you're going into a bend really quickly, don't lift off too much because the car will swap ends on you!

8. **Ford Fiesta ST Mk7** – A punchy 1.6-litre turbocharged engine combined with a highly playful chassis makes this a great affordable driver's car.

9. **Ford Focus RS Mk1** – Looks the part, goes well and has a slightly unruly character because of the front-wheel drive but with a turbocharged engine (with quite a bit of turbo lag and turbo boost when it comes in). It feels like you're fighting it but it makes you feel alive!

10. **Mercedes-AMG A45 S** – This thing has a bonkers hand-built engine, four-wheel drive grip, it looks good and it goes like stink. It's the ultimate mental hot hatch, if that's your thing.

Hypercars

Koenigsegg

I have drag-raced a few Koenigseggs and I'd met founder Christian von Koenigsegg briefly at motor shows, but we had no luck getting out to Sweden and testing one of their cars. And then, in May 2025, we managed to set something up for Carwow.

I travelled over to the factory and was given a tour by Christian, who's a really nice chap and a massive petrolhead. I loved that he turned up at the factory driving a first-generation MX-5, so we got off on the right foot, seeing as I owned a Mk1 MX-5 and still have a Mk2. It's not every day that you give consumer advice to a founder and CEO of a hypercar manufacturer but it turns out he owns another car that's on my driveway – the GR Yaris. He was wondering if he should get the new one because he wasn't sure it'd be all that different to the old one. I had the old one and now I have the new one, so I found myself in the fortunate position of being able to tell him how surprisingly different it is to the previous version, and yet still be bloody amazing. So who knows – Christian von Koenigsegg might end up with the new GR Yaris based on that chat.

What amazed me on that tour of the factory was how cutting edge their engineering is, bearing in mind the small scale of the operation. I saw this gearbox that, rather than using a single clutch, has individual clutches for each gear, which means it changes gears so quickly without a drop in torque. It's a Tony Stark kind of operation, with a remarkable genius inventing remarkable things. He's got a certain look about him, Christian von Koenigsegg, like he belongs in a film. He's got character and so do his cars. His sense of humour shines through with the names he gives their engines and components, like the 'light-speed transmission gearbox' in their Jesko Absolut. You can do that when you make practically everything yourselves.

There are other details about the company that are so cool, like the fact that the factory is based on an abandoned airfield formerly used by a Swedish air force squadron that featured a ghost symbol on their aircraft. And so Koenigsegg use lots of references and mottos to this squadron on their cars, on parts that no one's ever going to see. That's the level of engineering detail that goes on at Koenigsegg. No one is going to know that there's a message on a conrod, which connects the piston to the crankshaft. No one except Christian. Unlike many other car manufacturers I've visited, the accountants aren't in charge. This place is run by a guy who loves making stuff. It's such a cool place to be.

I liked Koenigsegg when they first came out because they were run by this plucky Swedish guy who loved cars and set out to build the world's fastest car. To be honest, I wasn't sure about the cars themselves until I went out there to spend some time with Christian and the team. Now, I'd love to own one because they somehow combine the artistry of Pagani, where things just look incredible, with the spectacular engineering of Bugatti. It gives you the best of both worlds.

Every Koenigsegg seems to break records. The Koenigsegg Agera RS was the fastest car in the world. Then the Koenigsegg Regera was the fastest car to go from 250mph and then back to zero. That time – 28.81 seconds – was only beaten by another

Koenigsegg, the Jesko Absolut. I got to drive that car in May 2025. It's rear-wheel drive and it did the quarter-mile in 8.9 seconds on a dusty runway. There's no space-age fan, not even four-wheel drive. And what's interesting is that unlike other cars, where the first 60mph is the highlight, it's afterwards which is even better, when this car just absolutely motors.

We asked to drag-race their Regera against their Jesko Absolut. One of the engineering guys there said: 'This is the first time we've done this here.' I was shocked. 'How have you not done this before?!' I asked. 'You've got the runway. You've got the cars!' The Regera is a weird car with its big twin-turbo V8 engine, three electric motors and single-speed gearbox sending power to the rear wheels only. For your first bit of acceleration, it's the electric motors driving most of it. The petrol engine, meanwhile, is always in sixth gear and it's only pulling a little bit, until you're going faster and then, when the revs are rising, the turbo kicks in. At that point, the petrol engine is powering it.

Christian tries all this different stuff, especially with crazy gearboxes. Their engineering system and design approach is unique. For example, they've got this special windscreen wiper to clean the very last bit of the windscreen and it's like something out of *Inspector Gadget*. Everything they create is cool. But Christian isn't some kind of limitlessly wealthy billionaire. The company's been close to going bust in the past. I was in Christian's office, and he's got loads of memorabilia including a hand-sized model of the first prototype car they made, which they took with them to presentations to drum up some investment. Someone stole the model when it was with a courier in the States. About fifteen years later, it ended up on eBay, advertised for not that much money. Christian was lucky that someone spotted it and told him, and he was reunited with it.

Christian's son was basically brought up in the factory and

now he works there as the head of design. Christian's wife is the COO of the company. It's a proper family business. It reminds me of my next door neighbour's family business, which had a little workshop making bits of metal for various automotive man-ufacturers. Just on a much bigger, grander scale.

All these fascinating figures who run hypercar brands are quite similar in the way that they love cars. They're very particu-lar about what they're doing and it's all about the product. They do make money and they've got an eye for business, but it's all about the product.

A bit like Ferrari, in order to be able to buy a Koenigsegg, you do have to build up a relationship with the company. Most people tend to buy a used one first before they buy a new one. I know a guy called Tom who creates computer games and has made a lot of money. He bought a used Koenigsegg first, and now he's buying his first new one. You're almost earning the right to buy one. We used Tom's Koenigsegg on the Carwow channel a few times and the videos did very well. That's why Koenigsegg took notice of us. We got a message from them telling us that Tom's car wasn't running on E85 fuel, which has a higher ethanol content and therefore allows you to deliver more power. So they got in touch to invite me to Sweden because they wanted to show us – with the proper fuel – what their car could do. And my god, they weren't wrong. Launching the Jesko Absolut was insane. I bloody loved it.

McMurtry

The McMurtry Spéirling (Spéirling meaning 'thunderstorm' in Irish) is an electric 1,000hp single-seater rocket ship of a car. It's crazy. The videos online don't even look real, like the one where the Spéirling overtakes a Porsche GT3 RS on the outside of a bend, or when it smashes the Goodwood Hillclimb record. It's the fastest-accelerating road car in history, yet it still looks like the videos are on fast forward.

This car is all about taking you for the track-day ride of a lifetime. Its unique fan-powered downforce system generates two tonnes of downforce, which basically sucks the car to the ground. Usually, with an electric car, the rapid acceleration is the big draw, but with the Spéirling, you also get the cornering capability of a Formula 1 car to go with it. Only, with a F1 car, you have to be travelling fast for the downforce to actually work, which is a leap of faith. You have to trust it. But in the McMurtry, you can just select maximum downforce, wait for the fan to crank up and stick your foot down. Motoring journalist and racing driver Chris Harris took the McMurtry Spéirling on to a customised rig that rotates all the way around. The McMurtry was upside down and yet, with those incredible fans providing that extraordinary amount of downforce, the car didn't move. No magnets, no magic show stuff, just astonishing engineering.

When I took it down the straight at Silverstone, the traction it got off the line for a rear-wheel drive car was nuts. I did 0–60 in 2.27 seconds, which was unbelievable, but I'd actually done faster in the Rimac Nevera. At this point, the McMurtry engineer told me that he'd only given me 650 of the 1,000hp on offer for the first run. On the second launch, he dialled it up. On my final run, the car did 0–60 in 1.4 seconds and the standing quarter-mile in 7.97 seconds. That was – unofficially – a world record. It's a catapult of a car. Obviously, I took the opportunity to do something silly with the fans so we went over a load of leaves and fired them out over the back of the car. It made me wonder what would happen if you took the car over gravel, because you'd effectively be firing a shotgun at everyone in the vicinity.

I do think that the fact it's a single-seater is a shame. You want it to be a two-seater so that you can take your friend in it and show them quite how crazy it is. Because after you've got used to the acceleration, it's fun to scare the crap out of someone else and witness their reaction.

A few months after launching the Spéirling at Silverstone, we convinced McMurtry and Rimac to come along for a drag race at the bottom of the runway we usually use. Rimac thought they'd win but their Nevera was struggling to get traction on that part of the runway because it's a bit slippier. There's a section of the runway further up, which has ridges and therefore more grip, so the Rimac guys asked if we could move up there. I was OK with that, but I asked the McMurtry guys, basically saying: 'Come on, guys, you're still going to win.' They agreed, but those little ridges in the track meant that the Spéirling didn't get full suction, so their launch wasn't quite as good as it could be. While the McMurtry beat the Rimac off the line, the Rimac eventually came past.

Part of the problem was that the Spéirling had hit its limiter (well, not a limiter as such, but the really short gearing it uses) at 150mph. With different gearing or maybe a two-stage gearbox, the Spéirling would have taken down the Rimac. It would have also beaten the Rimac if it hadn't been on a bumpy surface because, ultimately, it is the quicker car. Who knows – it may have even done the standing quarter-mile in sub-seven seconds, which would have been nuts. But ever since then, McMurtry have been slightly less interested in dealing with Carwow. I wonder if they're a bit mad at me for moving the start point. I get it – they're selling their car based on the performance and everyone saw that the Rimac beat it. To be fair, the difference in performance is very close. The Rimac does have a major advantage: you can just jump in and drive, but in the Spéirling, you have to turn on the fan and make sure you don't keep the fan on for too long when the car's stationary. You've also got to make sure there's no one behind it, all this kind of stuff. Yes, the McMurtry will go quicker than the Rimac on paper, but it's a bit of a hassle. Get it on a track, though, and the McMurtry will beat anything. It's an insane car.

The Spéirling is a plaything for thrill-seeking billionaires. If you've got loads of money and you want to experience Formula 1-like cornering ability, it will give you an experience like nothing else. If I was a billionaire, I'd have one. I'd live at a track, I'd have my own car museum there open to the public and visitors would be able to take cars out on the track. I'd be a benevolent billionaire, by and large. But I'd still be a petrolhead, so sometimes people would arrive at the track only to find that it's closed because I'm pissing about in my McMurtry. Someone asked me recently if I'd take a Rimac Nevera or a McMurtry. If I was a billionaire, I'd just have both, but I'd lean towards the McMurtry because it's the more unusual car, the fan adds an extra dimension and you really can scare the shit out of people in it, which is a lot of fun. But I'd take the Bugatti Tourbillon over both of them.

The Spéirling is the car above all others that has genuinely given me a 'holy f**k' moment when you put your foot down. Before you activate the fan and stick it into launch mode, the McMurtry guys tell you a number of times: 'Make sure you turn off the fan when the car's not moving.' 'Don't worry – I'll definitely do it,' I said at least five times, wondering why they seemed to think I was a moron. I did the run and it completely scrambled my brain, which I'd left 400 metres behind me. I couldn't remember how to put the car into reverse. I came back to where the engineers were with the fan still whirring. The McMurtry gave me amnesia. It rearranged my brain.

Pagani

In my twenty-plus years as a motoring journalist, I'd never driven a Pagani. To be fair, very few people have driven Paganis, but then, in May 2025, we got a message at Carwow inviting us to a very special event. It was at the Spa racing circuit in Belgium and I'd get to drive Pagani's brand-new track-only Huayra R Roadster. That car has a completely bespoke racing chassis, with an engine produced by the same people who made the engines for Mercedes' Formula 1 cars. So, the first Pagani I'd get to drive would be the newest and most exclusive Pagani they've produced. It was a baptism of fire, especially as I had to sign a disclaimer that if I crashed it and was seen to be negligent (I'm still not quite sure what that would have entailed), I'd be liable for the cost. We're talking at least £3 million. Fortunately, I didn't have to sign a physical copy of the disclaimer because I'm pretty sure my hands were shaking, so my signature would have looked a bit like Guy Fawkes' confession.

The day before I arrived, one of the Pagani owners span his Huayra R Evo on the famous Eau Rouge section of the Spa track but somehow or other, he managed to pirouette 360 degrees and not crash into anything, which was a miracle. The onboard footage, which you can see on Instagram, is completely mad. Spa is a fast and scary track at the best of times and it would be the first time I'd be driving on slick tyres so, as you can imagine, I was bricking it.

The first evening I was there was billed as a formal dinner, and I didn't really know what to expect. It could have been the full tuxedo on the *Titanic* kind of thing but I gambled and wore a shirt and dug out some smart shoes. I got there and everyone was dressed quite informally, but it's all somehow smart because everyone's super wealthy so everything fits perfectly, lots of one-of-a-kind cool trainers, that sort of thing. I was apprehensive that I was going to be completely out of place – a comparatively skint motoring journalist thrown in with some of the world's wealthiest people. But Christopher Pagani, the son of founder Horacio Pagani, was such a super-relaxed friendly kind of guy and he put me at ease. As it turned out, all the billionaire Pagani owners there were massive petrolheads, who kept telling me that they loved Carwow's videos. It was all very surreal.

The following morning, I was up early at the track, filming the car, interviewing the people who worked for Pagani, then eventually going out in the car. The event was part of Pagani's 'Arte in Pista' (Art of the Track) programme, which you access if you buy one of their cars. It involves Pagani hiring out a racing circuit and then the billionaire thrill-seeking Pagani owners from all over the world turn up in their private jets and get a Formula 1-style experience. Pagani look after the owners' vehicles for them, transport them to the track and supply customised helmets and safety equipment, with engineers and technicians on hand to fix anything that comes up. They've even got nutritionists, physiotherapists and massage therapists there to keep everyone in peak condition, hydration and energy-wise. It's an unbelievably exclusive set-up, basically having your own race team for a weekend. Pagani don't make any profit out of these events – it's all just part of what they do.

Pagani owners can use the track as much as they like at these events and there's an open pit lane. I was given four laps, which was probably a blessing, because more laps equals more risk equals rising likelihood of paying Pagani £3 million to fix the car. I had an amazing time with the car, although in that time, you have to acquaint yourself with a multimillion-pound 900hp hypercar, cope with the slick tyres, film it, hopefully entertain people and actually produce a decent time on the track so you don't look like a fool to your viewers and peers. We analysed my data afterwards and compared it to Pagani's development driver and I was a fair bit off the pace. But I didn't damage the car and need to remortgage my house, so that was something.

These kinds of experiences are intense and knackering. When you come back home after an event like that, your perception of what a fast car is has been rearranged. I remember spending a day drag-racing a couple of tuned Audi R8s, whose horsepower was north of 1200. The following day I drag-raced the standard Audi R8 and I was absolutely convinced it was broken.

When I came back from the Spa event with Pagani, I was wiped out for two days. It was the same with my other hypercar experiences. My first time driving a Rimac Nevera involved meeting Mate Rimac and then doing a drag race against him in a Ferrari SF90. My first time in a Bugatti Chiron was taking it on the autobahn to go as fast as I could with their legend of a development driver, Andy Wallace. The first time on a Koenigsegg event, I drag-raced their Jesko Absolut on their runway at 240 miles an hour. My first experience in a McMurtry was the quickest I've ever accelerated in a car. It's a bit like stepping into the boxing ring for the first time and you look up and realise that your opponent is Mike Tyson after he'd just come out of prison.

Rimac

In 2007, Mate Rimac was a nineteen-year-old who liked nothing more than drifting his 1984 BMW M3 E30 around a track. Then the engine blew up. Instead of fixing the engine, he started converting it into an EV. In his garage. All the pioneering business geniuses start off in a garage. That vehicle became the fastest-accelerating electric car in the world.

In 2009, Mate founded his company with a crystal-clear goal: *build the world's first electric hypercar.* He hired premises near Zagreb, Croatia (where's he from) and got to work. Just two years later, he launched the Concept One at the Frankfurt Motor Show, a car that blew a bunch of supercars away and got everyone talking. Only eight were produced and Richard Hammond famously crashed one of them after completing the Hemberg hill climb event in Switzerland while filming a *Grand Tour* episode in 2017. The car flipped and burst into flames but Richard was able to get out and survived with a broken leg. It could have been the end of the company but the publicity worked in their favour because the car's ability spoke for itself. Rimac went on to build a production car called the Nevera, and, in doing so, they became the leading experts in battery and motor technology in Europe. Now, they produce batteries for loads of European car manufacturers.

I met Mate Rimac when he was launching the Nevera in 2021. He's an extremely bright, friendly, low-key guy but he's got a ser-iously motivated side to him and clearly has the killer instinct you need to be in charge of a company like that. He moved out of Croatia during the war in the early nineties but came back. And he was very keen to build the factory in his homeland even though he didn't get any tax breaks from the Croatian govern-ment and he was being heavily encouraged by other governments to set up his factory in their countries. Above all, he's very keen to make sure that Croatia benefits.

When I met him, he told me he's a fan of Carwow, which was a pretty cool moment. And he brought a gift. To honour the 'stick of truth' thing I do to test whether or not a car has a fake exhaust, he'd made me a carbon-fibre stick of truth that vibrates.

I mentioned this on a video on my personal channel, Mat Watson Cars, and the video got demonetised by YouTube for featuring sexually explicit content.

Building a hypercar is a terrific opportunity to showcase what you can do, but it's not where Mate actually makes his money. That comes from his skills in battery tech. Rimac leads the whole of the European continent in that field. That's the hook that got Porsche interested in investing in Mate's company and buying 45 per cent of Bugatti Rimac in 2021. The batteries that go into Porsche's hybrid cars, like the 911 GTS, are made by Rimac.

Mate's a real petrolhead. Lined up outside the Rimac campus, on a big site outside of Zagreb, is his Porsche Carrera GT, his beloved M3 E30, an SLR McLaren and a Bugatti Chiron. They're the kind of cars that car fans appreciate. As a spin-off to the main operation, he's also created a self-driving car that they'll roll out in European cities in the next few years. That's his genius entrepreneurial side coming through, because at some point in the not-so-distant future, we're not going to be doing that much driving ourselves. Whatever the motoring world will bring us next, you can bet that Mate Rimac will be at the front of the grid.

Hyundai

What a turnaround.

When I was at school in Walsall in the West Midlands, my mate's mum had a second-generation, four-door Hyundai Pony. It was the first mass-produced car that had ever come out of South Korea. After they'd taken all that trouble, you'd think they'd have come up with a better name, given that 'pony' in Cockney rhyming slang means 'crap'. To be fair, they weren't aiming high at that point. They were pitching themselves above the likes of Lada and Skoda but not-quite-a-Nissan, which to my mind puts you in the *Pretty shit but not so embarrassing that your kids are going to get beaten up at school* category.

Working at *Auto Express* in the early noughties, if I was given a Hyundai to review, I knew that I must have pissed off our head of admin because she wielded the unparalleled power of allocating the cars to journalists. That made her more important than the managing editor in many ways, so we made sure we kept on the right side of her. If you were in her good graces, you'd end up with the Astons and Porsches, so she was well worth bribing with a pack of posh biscuits. Having said that, I was still quite junior, so the biscuits would only take you so far down the road. I remember the day when I was finally lobbed a key to a Renault Clio instead of a Kia or Hyundai and it felt like I'd been promoted. But that day was still a way off as I trudged over to whichever Hyundai I was given to review that day. *Same size as a Golf, not as good as a Golf, bought by people who can't afford a Golf.* To be fair,

I probably should have submitted more than that nineteen-word review to my editor.

Twenty years later we awarded the Carwow Car of the Year for 2025 to the new Hyundai Santa Fe. This big SUV is a high-quality, efficient, futuristic car that drives nicely and looks like the Land Rover Discovery should. It's hard to imagine writing a sentence like that twenty years ago.

In the mid-to-late noughties, Hyundai began the motoring equivalent of the *March of Progress* illustration that shows 25 million years of human evolution. The first thing Hyundai did was to headhunt talent from Europe, adopting a kind of 'Avengers assemble' recruitment strategy. Designer Thomas Bürkle was one of the top guys at BMW, and when he joined Hyundai in 2005, he started assembling the brand's DNA.

Peter Schreyer, who was well known for his work designing Audis in the late 1990s, including the TT, was recruited by Kia in 2006 as their chief design officer. He was largely responsible for reinventing Kia. Hyundai are the parent company of Kia, and in 2013, Schreyer became the chief designer for the wider company, Hyundai Motor Group. You could see what he was aiming for: *Get us as good as a Golf.* They weren't far off with the third-generation Hyundai i30 but there was no real desire for it. They needed some showstoppers, so who better to bring on board than the chief engineer of BMW's M division.

When Albert Biermann joined Hyundai in 2015, it was big news in the industry. He'd been at BMW for more than thirty years but the prospect of starting up a completely new high-performance division from scratch must have been a dream job. He created their 'N' brand and their first model, the i30N (released in 2017) hot hatch, brought Hyundai right up there in terms of fun, character and driving dynamics. Hyundai say that they named their performance division in honour of two places: their

Namyang R&D location in South Korea; and the legendary Nür-burgring, where Hyundai test and hone their performance models. But part of me reckons they went with 'N' just because it comes after 'M'.

After that, Hyundai focused on EVs and it's been their master-stroke. The Ioniq 5 N is magnificent. Introduced in 2023, it looks exactly like I imagined cars of the future would look when I was a kid. The front looks like a *Star Wars* Stormtrooper, the wheels look like something you'd try and draw on a Spirograph and the rear light bar is straight out of *Tron*. Tesla could learn a thing or two from Hyundai about exterior design.

It's also just as good inside. The Ioniq 5 N is the most fun-to-drive electric car because Hyundai introduced paddle shifters, which simulate an eight-speed dual-clutch gearbox. It sounded like a gimmick when they announced it but it's terrific. With regular electric cars, you put your foot down and get sudden acceleration, which tails off. What Hyundai do, in their 'gear mode' setting, is not to give you all the power at once. So, in 'first gear', they give you some power, and then when you pull the next gear, they give you some more power, but the increase is not so much. The car isn't actually as fast in gear mode as it is in normal mode, but gear mode feels faster because you're changing gears. Hyundai have given us the thrill that we're all missing from electric cars. They thought of it and went out and did it. None of the European carmakers have.

The inside of the Ioniq 5 N is quite minimalistic and futuris-tic but they've done without the Hyundai badge on the steering wheel, going instead with the funky 'N' logo. I think because the Ioniq range is unique and so different, they wanted to detach themselves from the possible prejudices that people might have had over Hyundai's previous product range. They wanted people to experience Ioniq for what it was without any of that baggage.

The Ioniq 5 N is a fair price bearing in mind the performance you're getting from it, the size of its batteries and the tech on board. But that's not a surprise to me. When it comes to EVs, particularly, Hyundai offer a top-end package at non-top-end prices. What's not to like about that?

Hyundai also give you a five-year warranty on their cars in the UK. Incredibly, it's ten years/100,000 miles in the US, which is a massive statement of confidence in their product and quality control. Although you do have to be careful with warranties in general and their tapering returns on what's covered, I can tell you that there ain't no British, German, French or Italian manufacturers dishing out warranties anywhere near that long.

People aren't choosing Hyundai because they're cheaper than European brands and do a decent job. People are saying 'I want that car'. On the EV front, Hyundai aren't just a bit ahead. They're embarrassing the European carmakers. They've mastered the art of combining European design flair with Asian quality and reliability, bringing them together in electric dreams.

What Hyundai aren't so good at mastering is my sense of humour, and something I said on my review for the Ioniq 6 pissed them off a tad. I may have mentioned that the drive-mode selector on the steering column bore a resemblance to a pocket sex toy.

Before then, I hadn't been invited to Hyundai HQ in South Korea, and I can't see that changing any time soon.

At Carwow, we still get sent cars by Hyundai, but I sense that there's a nervousness about what I'm going to say. I heard it on the grapevine that the Hyundai guys in South Korea watch my videos and enjoy them, but the reason that I don't get invited to South Korea to test their cars is that if I offend them when they've flown me out there, their bosses are going to give them grief.

I recently went to a motoring event in Germany, filming

drag-racing, and the Hyundai team from South Korea were there doing some work on a Genesis (Hyundai's luxury car division). Suddenly, all the members of the engineering team came over and were shaking my hand and asking me for selfies. In time, I hope their bosses will come round. To be fair, there's always a bit of wrangling between journalists and car manufacturers and occasional fallings-out, but you always make up in the end.

Jaguar

I had just started as a motoring journalist but the joy hadn't sunk in yet. The moment it did and I said to myself *Woah – this is actually happening!* was when I was sat in my publisher's S-Type and saw the Growler badge on the steering wheel. *Oh my god, I'm driving a Jaguar.* When I was a kid, Jaguar were *the* brand. My best mate's dad (who ended up becoming my dad's best mate too) had a few cool cars, like a BMW M535i, but his Jaguar XJ6 was the business. It was the one with dual fuel tanks and even the fuel caps looked amazing – they were chunky, chrome and with a thumb-shaped segment at the bottom that slid aside to reveal a key slot. Inside the car was that wonderful 'J-Gate' (J-shaped) automatic transmission, so you could select P, R, N and D on the right, or slide the gearstick down and across to choose the individual gears on the left.

When I was growing up in the early 1980s, the division of wealth wasn't what it is today. My dad was a teacher and my best mate's dad was the MD of a company but aside from his family going to America for a few holidays and having slightly fancier wallpaper, there didn't seem like much of a difference. We lived on near-enough the same street, a regular suburban road in the Midlands with three and four-bedroom detached houses, as did West Bromwich Albion right-back Brendon Batson. Everyone's kids went to the same school and the parents all hung out together. Today, the three families wouldn't live on the same street. The Premier League footballer would probably be in a gated estate and the MD might be in a manor house in the country, but my family would still be on the same street. The upshot is that back in the 1980s, I got to see nice cars on my street and up the road much more so than I would now. The richer people on our road, with a bit more disposable income, had Jaguars.

One of them, another mate of my dad's, owned an XJS, and I'd never seen a car like that before. Not only was it elegant and beautiful, it genuinely looked like a leaping jaguar. And those 'flying buttresses' that sweep down from the rear window to the rear of the wings not only looked amazing – they also reduced drag. The buttresses were actually vestiges of a mid-engine concept designed by Malcolm Sayer (who also designed the E-Type) but Jaguar ultimately abandoned that idea and went with the front engine.

The XJS left a big impression on me, to the extent that I remember saying to my dad: 'One day, when I'm old and rich, Dad, I'll buy you an XJS.' When my dad was in the final stages of cancer, I wasn't wealthy, so I took him out for a drive in a XK coupe that I'd borrowed. It seemed like the next best thing after an XJS. When I showed him the car, I said: 'Dad, I can't afford to buy you the XJS that I promised but I have been able to blag a Jag.' Looking back on it now, it reminds me of how the two of us were absolutely cut from the same cloth. My dad couldn't afford a Jag as a family car, so he arranged the next best thing – he managed to blag his mate's XJ6 to take us on holiday one year.

In the late 1990s, Jaguar developed a reputation for being a favourite among the pipe and slippers brigade. A lot of that had to do with the S-Type, which was 100 per cent an old man's car. That car always seemed to smell of cigarettes, even if the driver didn't smoke. Maybe Jaguar deliberately infused the cars with tobacco smoke so they smelt like a drawing room at a gentleman's club.

Porsche kept evolving the design of the 911 and did it in a

way that always looked modern, but Jaguar seemed to stick to their guns with a retro look and I think it affected their brand image. The third-generation XJ (the X350, produced from 2002–09) looked very retro but was actually more advanced than all of its competitors. It had a fully aluminium body, the engine was really up to date and it was high-tech. The trouble was it was still clothed in a dressing gown.

Designer Ian Callum took over after Geoff Lawson died in 1999 and completely ripped up the old design with the next generation of Jaguars, which started looking really modern. By that point, though, it was too late. Jaguar had two big hurdles to leap over: a fusty brand image and the fact they always seemed to be trying to compete with the Germans and not doing it quite so well. Jaguar have done better in the past by forging its own way. The E-Type and the D-Type were cool, innovative cars that were ahead of their time. But even with the second-generation XK, which came out in 2006, everyone said the same thing. It looks like an Aston Martin DB9, which had been out since 2004. The similarity may not be much of a surprise, seeing as Callum designed both. It's no bad thing looking like a DB9, but Jaguar were a brand built on the tagline 'copy nothing'. In the past, Jaguar had made their name by being the car manufacturer that was leaping forward.

It didn't matter that some of the proud claims about the E-Type turned out to be a bit of a lie, like the 150mph top speed or the sub-seven-second 0–60. The car that achieved those stats was a ringer, for sure. The performance of the regular production model was still 145mph, but no one was looking too closely at the stats anyway. They were all hypnotised by the look of the car – even Enzo Ferrari, who called it 'the most beautiful car in the world' at the 1961 Geneva Auto Show. My dad's mate had an E-Type coupé, and I always preferred that to the roadster.

Sure, it probably isn't as conventionally beautiful, but something about the slightly awkward shape and the weird back door really appealed to me. It wasn't just cool; it was also unusual.

The XJ220 seemed to come out of nowhere in 1992. Suddenly a British car manufacturer had made the fastest production car in the world. But there were a lot of failed promises. The concept car had a 'targeted' top speed of 220mph, so they called it the XJ220. It did eventually reach 217mph, after the exhaust catalysts were removed and the rev limiter was nudged up. The XJ220 was supposed to have a V12 engine, but it ended up with the substitute Austin Rover rally car engine, the MG Metro V64V. That's not a good look for a 'halo' car, especially when hundreds of people have already put down deposits. Yes, Jaguar did develop it on a shoestring budget, and for a year it was the fastest production car in the world, which was some achievement, but you just got that sense that the whole operation was held together with cardboard. Instead of working out if something was actually possible before making a promise, they'd make the promise and hope for the best. Professor Jim Randle, former director of engineering at Jaguar, summed it up perfectly in an online article, when he said: 'Selling the sizzle became more important than selling the sausage.'

The trouble with this reputation of overpromising and underdelivering is that, when a new concept car comes out from behind the red curtain, you do wonder if it might reveal the Wizard of Oz projecting a car using smoke and mirrors. Jaguar revealed their new electric concept car, the Type 00, in the first week of December 2024, and, with its letterbox headlights, angular front and square grille, it looked like something designed for a cartoon villain. Jaguar have thrown a Hail Mary; or, to use a more British metaphor, it's an injury-time hoof up field and even the keeper's gone up to try his luck. To be fair, Jaguar had warned us that something ridiculous was on its way after dropping a 30-second rebrand video a couple of weeks before, which looked like a cross between a perfume commercial, a Tate Modern installation and a new soft-play arena at IKEA. One thing conspicuously absent from it was a car.

The insurmountable problem with Jaguar is that no one cares any more. Us motoring journalists do but the buying public don't. They've had too long as also-rans to the big German manufacturers. It's a shame because their first foray into the EV market, the I-Pace, in 2018, was a special car. Sporty, sleek, fast and fun to drive, with their most modern interior to that point, Jaguar were back to what they were best at – leading from the front. The problem was that it proved to be a bit of a blitzkrieg car. It blew everyone away at first, but they had no plan to hold the ground they'd seized. Where was the next wave – the second generation?

The 00 looks mad. So did the (electric) Rolls-Royce Spectre, and it's depreciating harder than petrol-powered Rolls-Royces. I was filming in a Dubai car dealership in February 2025 and they had at least £150 million worth of cars in there. I asked the manager which brands were depreciating fastest. 'The English ones. It's always the English ones,' he said. I asked him to name a specific model that, depreciation-wise, gives him the shivers. 'Spectre,' he said.

Looking at the window line of the Type 00, it's going to give the driver a terrible view and I'm not sure they're actually going to be able to do the front end, because I doubt it will meet crash-protection regulations unless it's softened. That's the reason the Tesla Cybertruck won't be coming to the UK or Europe – because the sharp edges will chop someone in half. Oh, and with four passengers and luggage, it would be over the 3.5-tonne weight limit allowed as part of a standard Category B driving licence. And the 00 does feel a bit like the Cybertruck, in the sense that it will attract a lot of interest and may sell well initially, but then people will start thinking: *Actually it's a bit pants.* No private buyer will buy it with their own money unless it's some crypto bro wanting to stand out. It will be on some kind of company-car scheme where you can make car choices without worrying about the depreciation.

The marketing strategy for the 00 has been divisive, but I guess that's to be expected when you ditch the pipe-and-slippers, patriarchal-olde-worlde-Britain vibe and are teleported to a Dolce & Gabbana catwalk show. One thing I did enjoy about the launch, at Paris Fashion Week, was that they had to charge the 00 using a diesel generator. The 00 has certainly got people talking. But for all the razzle-dazzle, you have to ask: who's it aimed at, apart from possibly Cruella de Vil? As for Jaguar, it feels like they've spun the roulette wheel for what could be the final time. They're placing everything on the 00.

What Jaguar could have done is move with the times, build some good-looking, reliable cars and sort out their organisational structure and corporate direction. Jaguar are still planning to go all-electric in 2025, but other major manufacturers, like Porsche, Ford, Mercedes and Volvo, are rolling back on their electric plans because they believe that the EU's zero tailpipe emissions by 2035 target will be pushed back. Ultimately, we live in a capitalist country and if the company's inefficient or you don't produce things that people want, you're not going to go the distance. It's life, and Jaguar's life, I think, is coming to an end. I hope that it works out for them, but my gut tells me it won't.

JLR (Jaguar Land Rover) reported a pre-tax profit of £2.5 billion in May 2025 – that's 15 per cent up on last year and their highest profits in a decade. But it's all down to Land Rover. The Defender alone sold more than 115,000 units; the Range Rover Sport nearly 80,000 units. In that time, Jaguar sold 26,862 cars. Total.

Their new-direction advert in 2024 succeeded in getting attention, but not the kind they wanted. But that's going to be the consequence if you alienate your base, while trying to appeal to a fashionista demographic that isn't going to buy your cars. Jaguar fired their marketing agency, but no one's going to be able to change the main issue: the Type 00 car itself. No one is interested in expensive luxury EVs. Maybe if they'd done a big, powerful hybrid or adopted Bugatti's strategy of producing an ultra-high-end car in limited numbers aimed squarely at the super wealthy, they would have been on to a winner.

JLR can take the hit because Land Rover and Range Rover sales keep pumping, but you do wonder: is Jaguar worth retaining? Or should JLR just double down on the vehicles that people actually want?

Kia

When you're starting out as a motoring journalist, you get sent on trips to go and check out cars that senior journalists don't care about. It's the car journalist equivalent of making the tea or doing the photocopying, which I also did as a junior. One of those 'adventures' involved the first-generation Kia Picanto back in 2003. It was the budget end of the budget car market. It had no desirability whatsoever, and the only thing going for it was that it was cheap and didn't fall over going around corners.

Nevertheless, Kia's then UK MD was full of optimism, telling a room full of bored journalists that in five to ten years' time, they'd be competing with Volkswagen. No one believed him. To be honest, no one really cared. Some of the journalists seemed far more excited about lunch.

Six months after the launch, Kia were building a €1.7 billion car factory in Eastern Europe. They were serious. In 2003, they flogged just over 100,000 cars in Europe. By 2014, it was over 350,000. So what happened?

The game-changer was the second facelift to the second-generation Sportage in 2008. That Kia offering suddenly looked like a decent car and the data was looking pretty decent too. I was even prepared to put someone else's money where my mouth was because I recommended it to my then girlfriend's stepfather, who bought one based on my advice and was very happy with it. That second-gen Sportage had a good diesel engine, was solid value for money, looked pretty smart and did all that you wanted it to do. Kia turned a corner with that model. Yes, people bought them primarily for their affordability, but they were pleasantly surprised with what the car offered them for that money. Like for like, they'd probably go for a Volkswagen or a Ford, but the prices weren't like for like.

While it took a little longer than ten years for Kia to truly compete with the likes of VW, it's happened. It was the same in the mobile phone market with Samsung, a fellow South Korean manufacturer. No one thought they'd be able to compete with Apple in Europe. But as tech has become a bigger and bigger feature of cars, the Korean manufacturers have got bigger and bigger. They've beaten the Europeans in a tech drag race, and they've pulled level with them in terms of car dynamics, interior and exterior design. And they come with a class-leading seven-year warranty, a measure of how confident they are with their cars.

Kia have been on a steady march upmarket underneath our noses. It used to be the Japanese car manufacturers that European carmakers feared. Now it's the South Koreans. And I can guarantee that a fair few of us uttered the sentence: 'Shit, is that a Kia?' at their rear-wheel drive saloon, the Kia Stinger, in 2022. Sure, it flopped in the UK, and even though it had a V6 engine, the engine and gearbox combo just didn't match up to a BMW. They were trying to beat BMW and Audi at what they've been doing for years. But they knew they weren't on to a winner, so they side-stepped and started playing to their strengths: tech. Entering the EV market just felt like a natural, authentic move for Kia and it's proving to be a masterstroke.

In internal combustion engine cars, the most expensive part is the engine and the gearbox. With electric cars, the most expensive bit is the battery. Much of the quality of an EV basically comes down to the battery and the software needed to manage power delivery and energy conservation, and Kia have been leading the charge. The European carmakers haven't been able to compete on battery tech until fairly recently. What they specialised in was constructing good engines and good gearboxes and making them work together in harmony. Plus, they possessed the craftsmanship and design flair to put it all together in a stylish package. But now Kia are on top of that as well.

Kia's boldest car to date is the EV9. Back when I was looking at the Picanto, the EV9 would have been unthinkable. Laughable even. But, it turns out, this was the future Kia were dreaming about. And they've only gone and made it happen. The EV9 is a really cool, huge seven-seater SUV that looks like it's been drawn up for a cyberpunk film. It's got a stylish dash with a digital display that controls most of the car's features, but you do have some shortcut buttons for the temperature control. Thank the lord. You don't want to be leaning around a steering wheel and scrolling to change the temperature while you're driving. You want it to happen now.

And it's just got lots of cool little features. There are these springy, trampoline-like headrests; a glove box that's so big it's more of a glove garage; funky, chunky wheel arches; and wheels that look like something from a futuristic sci-fi film. The vanity mirror is basically a full-length mirror and when you turn the car off, it sounds like you've got your own theme music. The rear seats are as comfy as the front, so you don't feel like you're slumming it in the cheap seats, and they recline. They even swivel around 360 degrees. Then you've got features like when you indicate, it'll show your blind spot on the dash. It's got the tech you want, plus tech you didn't know you wanted, and it all integrates perfectly with your phone. It's spacious, it's comfortable, and it's super-easy to drive, with software that just helps you out. The EV9 has even got this function where you can park it using your key, transforming it into a massive remote-control car.

Kia's gone up in the world, and so have their prices. The entry-level version of the EV9 will set you back £65k, and it's up to £77k for the top-of-the-range model. Despite this, it's way better value for money than its competitors and looks really cool. Design drives desirability: that's why designers get paid more than engineers. If I told you the Kia EV9 did 0–60 in 4.94 seconds (on a horrible wet day), you wouldn't believe me. Thankfully, I filmed it. That's quicker than my Porsche 911 996. Bonkers but brilliant. So what's the sticking point? Well, some people are still saying, 'But it's a Kia.' The trouble is, no one's listening to them any more.

Lamborghini

The first Lamborghini I saw was the Countach from the opening of the 1981 film *The Cannonball Run*, which ticked every box. Action, check. Comedy, check. Unbelievable cars, check. Farah Fawcett in a jumpsuit, check. Every teenage boy's fantasy was playing out on screen. The Countach, which by the way is Italian for 'Wow!', didn't originally have a spoiler on the back, but in *The Cannonball Run* the film production team put one on the back and on the front to look cool. It worked. And because everyone loved that film, Lamborghini decided to add a spoiler on the back of the Countach, even though it caused drag and provided absolutely no downforce. But who cared when it looked like a spaceship.

Ferruccio Lamborghini made shedloads of money in the tractor business after the Second World War. So, in 1958, he bought a Ferrari 250 GT. This is where the tale starts to veer into myth, but in the interests of not letting the truth get in the way of a good story, here's what happened next:

Ferruccio bought a few more Ferraris over the next five years, but kept experiencing problems with the clutches, which would require a trip to Maranello to have them fixed. So, in 1963, rather than travelling back to the Maranello factory, he turned up at Enzo Ferrari's door, offering some clutch-specific constructive criticism. The legend goes that the meeting lasted about as long as a drag race, with a furious Enzo responding with:

'You may be able to drive a tractor, but you will never be able to handle a Ferrari properly.'

As Lamborghini later reminisced: 'This was the point where I finally decided to make a perfect car.'

So Ferruccio Lamborghini founded Automobili Lamborghini and just a few months later, he'd built a factory on farmland near Sant'Agata Bolognese, only twenty miles from Maranello. The legend goes that Lamborghini used to test their cars on the road outside Ferrari HQ in Maranello just to wind them up, which does sound plausible. One thing's for sure, Ferruccio wasn't hanging about. Lamborghini's first car, the 350 GT (a two-door coupé with a 3.5-litre V12 engine), was being produced from May 1964. It's quite a weird-looking car, mainly because of the headlights, which bear an unfortunate resemblance to a bar of soap.

The second Lamborghini design was a different animal. That was the Miura, which a lot of people say is the best-looking car ever made. Designed by Marcello Gandini at the hugely respected Italian industrial design company Bertone, you might remember it as the beautiful orange sports car being driven through the Alps in the opening of *The Italian Job*. It goes into a tunnel and becomes a fireball, before a thankfully pre-crashed Miura is collected by a bulldozer and deposited off the side of a cliff. Enter Mafia boss Altabani and his funeral-suit-wearing henchmen, who remove their hats before Altabani rolls a wreath down the cliff like a tyre into the river at the bottom. It's an iconic car, in an iconic scene in an iconic film.

Icon is such an overused word, especially in the previous sentence, but it's hard to find another superlative to describe the Miura. It was the first mid-engined Lamborghini, with the 3.9-litre V12 mounted behind the cockpit and in front of the rear wheels. The engine was designed by Giotto Bizzarrini and acquiring his services must have tasted mighty sweet for Signor Lamborghini. Bizzarrini was one of the 'Famous Five' engineers at

Ferrari and the man behind the stunning Ferrari 250 GTO, but he left his job in spectacular fashion.

In what sounds like a sub-plot from *The Godfather* trilogy, eight of the most senior employees at Ferrari, including the Famous Five engineers, staged a walkout. The reasons were complicated. The short version is that Enzo Ferrari's wife, Laura, was getting involved in the company's day-to-day operations. The longer, more nuanced version is that Enzo had become a withdrawn figure after the death of his son Dino in 1956 and Laura stepped up to lead the company, taking over his role of shouting at employees and doing a lot of frantic arm-waving. So the eight senior employees sought legal help and put together and signed a letter of complaint. Enzo's reaction is unclear, which is probably the best way of putting it because I'm not sure we'd be able to print what actually happened, but all eight of the senior employees left the company very soon afterwards.

For a man like Ferruccio Lamborghini, who had a score to settle with Ferrari, seeing Giotto Bizzarrini walk into the Lamborghini factory soon afterwards must have felt like the moment when Achilles agreed to fight for the Greeks against the Trojans. As for the Miura, it was the fastest production car in the world, it basically set the blueprint for mid-engined, rear-wheel drive performance cars, and, to put the icing on the panettone, the media crowned it the first 'supercar'. Never have two fingers been so successfully raised to a rival.

Incredibly, that flagship Bizzarrini V12 engine was the beating heart of Lamborghinis from the 350 GT all the way to the Murcielago, more than fifty years later. The engine was in its third generation by the noughties and had been extensively modified and improved, but the DNA's the same. As for the exterior of a Lamborghini, they do have a very distinct style. Everything's angular, edgy and pointy, a bit like a Transformer. There's no mistaking a Lamborghini.

I do enjoy Lamborghini's approach to naming their cars and I love what it tells you about the difference between Italians and Germans. Porsche's sports car names are numbers and they weren't even the numbers they wanted in the first place. Porsche wanted to call what became the 911 the 901, but Peugeot basically claimed they owned the legal rights to every three-digit number with a zero in the middle, so rather than go back to the drawing board, Porsche just shrugged their shoulders and changed the 0 to a 1. Pragmatic, efficient, sensible and swift. By contrast, Lamborghinis are named after fighting bulls. The Murcielago (which is Spanish for 'bat') was named after a bull who fought so valiantly at the bullring in Córdoba, Spain in 1879 that the crowd called for his life to be spared. And so he was presented to bull breeder Antonio Miura (which is where the Lamborghini Miura gets its name). Murcielago the bull had quite the after-party: he was put in a field with seventy cows to impregnate. No one can say that Italians don't know how to celebrate.

In 2013, I was invited to the Lamborghini fiftieth anniversary grand tour, which was completely mad. I had one of those phone calls from the UK press office that you can't really imagine as an eight-year-old when you're being hypnotised by the Countach from *The Cannonball Run*. It went like this:

'Which car would you like, Mat?'

'I can choose any Lamborghini?'

'From our current range, yeah.'

'I'll take the Aventador Roadster.'

'No problem.'

Ridiculous.

And so we embarked on a 750-mile, six-day drive in a two-and-a-half-mile-long convoy of Lamborghinis from Milan to the company's HQ in Sant'Agata Bolognese. We even had a motor-cycle-mounted police escort. When we stopped for lunch some-where in the stunning countryside, I asked a police officer, who was admiring the convoy in the sunshine with his aviator Ray-Bans and cigarette, what would happen if we went over the speed limit. He smiled, lowered his sunglasses and said: 'This is not the California Highway Patrol. You go as fast as you like.'

When we arrived at Sant'Agata, everyone was out on the street waving flags and cheering. That's how proud the Italians are of their cars. The sound of that convoy, with each one of us revving our engines, was unreal. It was before petrol particulate filters had been introduced, so the cars sounded even more incredible, screaming through medieval streets and being treated like con-quering heroes. It was an awesome thing to be part of. This kind of event wouldn't happen in the UK. There certainly wouldn't be any cheering. Maybe some mooning.

And that's the perfect note to bring in my friend Yianni Charalambous, because he and I met because of Lamborghini. Or perhaps that should be *in spite of* Lamborghini. It started when I wanted to do a drag race with a Lamborghini, but they weren't keen on lending me a car.

I already knew of Yianni through a guy that I'd shot a TV show with in 2013. He was already YouTube-famous for wrapping KSI's and various Premier League footballers' cars, and at that point, Carwow was starting to do some good numbers on YouTube. It felt like we might both be able to benefit from producing some content together.

Yianni's got this strange ability to get cash to stick to him. When I was still driving a shitty Ford Fiesta, he was in a Porsche 996. But I knew that the brand that Yianni loved above all others was Lamborghini, so I asked him if he'd lend us one. And so we started a series of drag-race bouts. In the blue corner was me in a Tesla; in the ridiculous, iridescent green corner was Yianni in a Lambo.

First up was the Aventador S against a Tesla Model S. The Tesla was so far ahead it was ridiculous. We did happen to have a Tesla Model X P100D on site as well and I figured that this heavy seven-seater bridge on wheels could run this V12 Italian thoroughbred close, which would make for more entertaining content. Yianni took a fair bit of persuading, though! Of course, on the first drag race, I pulled out the old playground classic *Stay rooted to the spot while your opponent starts running*. Yianni's reaction was perfect, getting all excited that he couldn't see me in his rear-view mirror. I figured this childishness was the best way to start our friendship.

When I uploaded the video, instead of doing a big spike on the first day and then trailing off, it just kept going like a Toyota Land Cruiser. Every day it was doing the same incredible numbers. It notched up 22 million views, becoming the second most viewed on the Carwow channel. And so a partnership of pissing about was born.

What you don't see on that video is that I managed to break Yianni's car. We'd mounted a camera on the windscreen and then a member of our team cleaned the windscreen, which triggered the wipers to start wiping. They got caught on the camera and couldn't go through their full rotation, which shredded the connecting bolt. I was shitting myself because we basically had no budget back then and I knew very well that if anything goes wrong with a Lamborghini, it'll cost five grand before you've even parked the car at the garage. Fortunately, Yianni wasn't too bothered about either the bolt shredding or me driving his car. I wouldn't have been quite so cool about it!

After losing the first Lambo vs Tesla drag race, Yianni whooped my ass in the second, with me in the Model X P100D. So

I went round the corner to find my bigger mate, the Tesla Model S P100D, to give him a beating. Yianni had a bad start but then annoyingly came through in the last few seconds in what I think remains one of the proudest moments of his life. That video also went viral and became one of the most-watched on Carwow's YouTube channel.

Next up was the Tesla Model S Performance against the Aventador S Roadster. Yianni was whingeing about traction but I couldn't hear him. I was busy winning and enjoying the feeling that while I was braking, energy was flowing back into the battery. Yianni won the rolling start race and the brake test and we finished 2–2, but to use Yianni's ill-fated words against him: 'It's *all* about the drag race.'

The Aventador (produced 2012–2022) is flawed but has loads of personality, and that goes a long way. On paper, its smaller brother, the Huracán (produced 2014–2024), is arguably the better car – it's got a brilliant V10 engine and a dual-clutch gearbox enabling seamless shifts that allow for perfect launches. It's also easier to drive fast, but it doesn't have the flamboyance of the Aventador with its mad V12 engine and carbon-fibre body panels. Somehow, the Aventador's jerky single-clutch gearbox just adds to its unique appeal. Of the two, it's the one that gives you the feels.

The Revuelto, the plug-in hybrid Lamborghini they unveiled in March 2023, is better in almost every single way than the Aventador, except for the fact it's a lot heavier, but it just doesn't hit you the same. They stuck a dual-clutch automatic gearbox in it but, even though it's a faster car, it doesn't give you that pleasing whack in the back the Aventador does. The Revuelto is also just a bit too complex and confusing. If you don't press this button and that one, the car won't be as quick as it can be. That's what happened when Yianni drag-raced it – the car underperformed for several races basically because he hadn't pressed the right combination of buttons in the right order. And that's with a person who's not exactly new to supercars. Lord help you if you're sitting in the car for the first time. It's also lost some rawness and personality and become a bit more homogenous and that's not what you want when you're in for half a million quid. You could have the very best Aventador for that – an SVJ (Super Veloce Jota, or 'Super Fast' J, 'J' being the FIA rulebook category for road-based race cars) – and you'll be happier for it because, while it doesn't have as much tech and isn't quite as plush inside, it's simpler, feels better to drive and gives you more thrills. Plus, while the Revuelto is dropping in price, SVJs are appreciating, so you can have some fun with a cool car and it'll actually earn you cash!

The Urus, Lamborghini's luxury SUV, first produced in 2017, is a great-selling car and holds its value well, but you could just buy yourself an Audi RS Q8 or a Porsche Cayenne if you want something that isn't so in-your-face and won't make you look as much of a dick. Motoring journalists are all very concerned about not looking like a knob. Sometimes I agree. But when I had the Porsche 911 GT3 RS, a lot of motoring journalists said: 'Great car, handles brilliantly, but you look like a prick, Mat.' I was thinking, *Have you not seen my videos?* I'm on YouTube dicking about every week. Why would I care? I like the look of it. I like the loudness. I like the fact that it's the kind of car you'd draw when you were a kid. I get the appeal of a Lambo.

But underneath, the Lamborghini Urus, the Audi RS Q8 and the Porsche Cayenne are the same car. It's a VW Group product. With the Urus, Lamborghini put a frock on it, tuned the engine a bit and made it feel like a Lambo. They absolutely nail their brand identity. The result is a noticeably different car to drive than the Porsche and the Audi. It's the same engine with a tiny bit more horsepower and speed. Of the three, the Urus has the biggest numbers, followed by the Porsche, then the Audi. But drag-race them all and who wins? The Porsche.

It's all down to the different approaches each manufacturer takes. I imagine the VW board of directors summoning in the Lamborghini guys to the boardroom to answer the following question:

'So, what do you want, guys, the most horsepower or to be the fastest?'

They'd go with horsepower. A Lamborghini buyer isn't going to notice the tenth of a second difference in a statistic that appears halfway down the page in the car spec. They're going to see a bigger number next to horsepower higher up on the page. Lamborghini owners are show-ers rather than growers.

Porsche will want to be the fastest. They know that Porsche owners don't care about horsepower, because Porsche will be working on deploying its power as well as it can and maximising traction to produce a quicker and dynamically superior car. It won't look like it's faster than the Lamborghini, but it will be faster.

The Urus is an absolute show-off car but it's cool at the same time. It's striking to look at. The design is edgy and interesting. In the *Who can make a box look sporty* stakes, they've smashed it and it looks visually more appealing than the dull-looking Cayenne. Inside, it's the same kind of thing. I thought the Audi RS Q8 was more comfortable than the Urus, but we did a vibration test on both of them and it turns out that the Lamborghini was slightly smoother. I just assumed that it wouldn't be, given the look of it and the noise that it makes, but the data told me that it was. Audi and Porsche have put a soft limit on their rev counters so you can't rev past 4000rpm when the car's stationary, partly to protect the engine, partly for noise regulations and partly for emissions. Lamborghini weren't having any of that rubbish. People might want to rev their engines outside a hotel to show off to their mates! They've paid for a bloody Lamborghini not a Nissan Leaf.

I can imagine the guys at Porsche HQ seeing a Urus rev up to show off and saying to each other (best imagined in a German accent): 'But why would they do this? There is no performance gain to be had!'

That's not the point of a Lambo. Each supercar manufacturer has their own vibe and appeal, but here's my general rule:

If you want a car to actually drive, buy a Porsche.

If you want a car for driving and showing off, buy a Ferrari.

If want a car purely for showing off, get yourself a Lamborghini.

Land Rover

Motoring journalists love Land Rovers and Range Rovers. They tend to score highly in tests because they do everything. They're spacious, they're comfortable, they drive really well both off-road and on-road, and they manage to somehow be both luxurious and not too showy. It's no surprise that Queen Elizabeth II loved both Land Rovers and Range Rovers and owned a whole fleet of them, including the Defender 110, the Range Rover Classic and the third-generation Range Rover L322. Ten of the Land Rovers and Range Rovers she used were displayed as part of a tour of North America in 2024, stopping at four locations across the US. Prince Philip even had a Land Rover Defender TD5 modified to carry his coffin.

However, the reason why I've never owned a Land Rover or Range Rover is because everyone I've ever known who's had one has had problems with it. It might be the electrics, the suspension, the engine or the gearbox, or sometimes a full-house of all four! These kinds of reliability issues are well-documented to the extent that Land Rovers have sat at the very bottom of the *What Car?* reliability survey in the past.

To me, Land Rovers and Range Rovers are a bit like smokers. The vast majority of them end up in very bad health, but once in a while you get one that confounds every expectation. It's your great uncle who's been smoking forty a day and he's in perfect health in his nineties. It makes no sense.

There's no denying that the more modern ones are hugely

desirable cars. When they're new, Range Rovers are bloody lovely. They look great. They're terrific to drive and they give you a very special feeling when you're at the wheel. And that feeling's the same whether you're a billionaire or someone who's saved up or stretched themselves to be able to afford one. They've got this regal, aristocratic feel about them while you're sitting above everybody else on the road, although, ironically, not higher than a Transit van.

Range Rovers were originally pitched squarely at the land-owning gentry. A 1974 *What Car?* review went like this: 'One feels that it has almost come to the stage now where no country house worth its salt is without one.' Christ, two 'ones' in a single sentence. Things began to change when the third-generation Range Rover came out, in 2001, when the company was under the ownership of BMW. But it was Gerry McGovern who really marched the brand upmarket with the fourth-generation L405 in 2012, the Evoque (2012) and the second-generation Range Rover Sport (2013). In that period, they'd detached themselves from country-house stuffiness and managed to become more elegant, refined and luxurious – but also more mainstream. It earned them a much wider customer base. And if anything, the prestige of the brand just keeps rising.

In 2019, McGovern absolutely nailed the new Defender. The old Defender was a utilitarian vehicle – quite basic in principle, but good off-road and relatively simple to fix. It had this niche appeal among guys in the City who liked the look of them but didn't want something that was crap to drive. And the old Defender was crap to drive. You were constantly correcting the steering like you were trying to control an untrained blind dog on a leash. Plus, the interior of the old Defender was just so out of date.

Even with all these problems, it was still an iconic car, so Gerry had some task in replacing it. With the new Defender, he managed to give a nod to the heritage of the original, while propelling the brand into the future. It drives really well and, because of all the tech they've added, it has even more off-road capability. What you wouldn't want to do, though, is go into the wilderness with it without support, because while in certain off-road stages, it's brilliant, if you knock a sensor off or if the new tech plays up, you wouldn't be able to fix it at the roadside with a set of spanners. You never had any of that to worry about with the old Defender when you were, I don't know, carrying sheep around or something. In that sense, the new Defender is a bit like a Rolex watch, which has painstakingly been put together by a human, but it's actually no more accurate than a Casio. Still, for what everyone's actually going to use the Defender for – not off-roading – it's a brilliant car.

One of the best press trips I've had was with Land Rover. To celebrate the sixtieth anniversary of the Land Rover, after Indian multinational Tata Motors had bought the company in June 2008, they put on a road trip along the ancient Silk Road – from Solihull, the home of Land Rover, all the way to Mumbai. They invited five motoring journalists and flew them in to drive different sections. I got the Delhi to Mumbai route via incredible places like Jaipur and Udaipur. It was all expertly planned and we had a full support team with instructors and medics. We got to stay at the Taj Lake Palace in Udaipur, the incredible island hotel they used in the Bond film *Octopussy*. The whole experience of driving through India was intense and there are very few cars that I would have liked to be in other than a very capable, comfortable, safe, easy-to-drive Range Rover. You will have seen clips of the traffic and the horns blaring in India but nothing prepares you for being in the middle of it. It's utter, utter chaos. The closest you get to it in the UK is trying to walk along Oxford Street on the Saturday afternoon before Christmas.

In one monumental jam, I took a turn and ended up going the wrong way over a bridge, weaving in and out of trucks coming in the other direction. I think that's the closest I've come to death. But I realised that it wasn't an unusual experience – it's just what happens if a lorry misses a junction, because the driver will do a U-turn and drive the wrong way down a motorway. In Britain you'll be on the news. In India, they don't even flash their lights at you.

In the UK, a cow has this worried look about it that communicates 'I'm a prey animal'. In India, because cows are revered, they look completely chilled, wandering happily across roads not giving a toss about anything. So as a motorist, you spend your time chicaning cows. A few times in the countryside, I saw drivers pulling over at the side of the road, getting out of their vehicles and sleeping on the hard shoulder. Instead of putting up a warning triangle, they put a collection of stones about 15 metres away from where they're lying down to 'warn' people. But somehow it all seems to work.

When you've been driving a long way, you've often been snacking a lot, so your car's full of rubbish and you just want to get rid of it. I stopped at a petrol station and grabbed armfuls of wrappers, packets and cans to put in a bin. Only there was no bin anywhere. So I asked the forecourt attendant and he lifted both arms up to indicate 'chuck it here'. It was the weirdest and most liberating feeling, standing in the middle of a petrol station forecourt and lobbing rubbish in all directions, and then being given a thumbs up by the bloke who worked there. I was so bemused by all this that I wasn't looking where I was going and stuck one foot into the completely wide-open trap door leading 20 feet down to the underground fuel tanks. That was the second closest I'd come to death on that trip.

Tata had their media guys out there documenting the whole

event and there was this one time-lapse shot they wanted to get of the Range Rover by this absolutely beautiful lake with the sun setting behind. When we looked at the footage of the sun melting into the car, it was spectacular. India will amaze you. In many ways.

Lotus

You might be familiar with the joke.

What does Lotus stand for?

Lots of trouble, usually serious.

Lotus's last great car was the Elise, which came out in 1996 and was only discontinued in 2021. The original Lotus Elan from the 1960s and seventies was cool. And it was around at a time when this little company from Hethel, Norfolk, were dominating motorsport. Founder Colin Chapman had a new approach. When everyone else was bulking up on horsepower he was doing the opposite, famously saying: 'Adding power makes you faster on the straights; subtracting weight makes you faster everywhere.' His Lotus 25 was the first Formula 1 car with a fully stressed monocoque chassis. It was structurally stronger than the other F1 cars and lighter, and driver Jim Clark didn't just win the 1963 World Championship – he obliterated the competition, winning seven of the ten races in the season. It was a similar story two years later in the Lotus 33. From 1963 to 1978, Lotus won seven World Championships. They were the team to beat. Also, they looked the coolest. When I was a kid, the best-looking F1 car was the black and gold of the John Player Special Lotus. I even encouraged my dad to smoke John Player fags just because I loved the livery of the racing car. Now that's a successful advertising campaign.

I've enjoyed driving Lotuses because they've always been fun. What wasn't so fun was the ball-ache of a drive to Norfolk when you needed to review a Lotus. Norfolk looks like it's *just there* on a map and yet somehow getting there takes about the same time as a transatlantic flight. Their HQ is a funny place – it's all woods and farms and then there's a little brown road sign for Lotus Cars. But that place is cool not least because you get to drive their cars on their very own test track.

I remember doing a video with Roger Becker, an industry legend of a chassis engineer, who worked at Lotus for forty-three years before he handed over the reins to his son, Matt, who took charge of the chassis development on the Evora (launched in 2008). A lot of chassis engineers spend their entire lives in a warehouse. One place you definitely won't find them is in the spotlight, but Roger was different. He was also an accomplished test driver. And these skills came in very handy in 1977 when James Bond traded in his Aston Martin DB5 for something sportier.

The Lotus Esprit S1 appeared in two Bond films: *The Spy Who Loved Me* in 1977 and *For Your Eyes Only* in 1982. Part of the first was being filmed in Sardinia and Roger Becker was tasked with driving the Esprit from Norfolk and then acting as a technical adviser, to help the stunt man, who as it turns out wasn't faring well in the Lotus. The story goes that they'd finished filming in one location and needed to move the car to higher ground. Up steps Roger, who took the car up the windy Italian hill mostly on full-opposite lock with tyres squealing. Director John Glen witnessed this with his mouth wide open. 'You're hired.' Those incredible scenes with the Lotus (well, except for the submarine bit) in *The Spy Who Loved Me* are all Roger.

The Evora, which launched in 2008 at the British international Motor Show, is a really good car. Lotus are famous for making cars that are brilliant to drive. They go over bumps so smoothly and they handle really well in corners. Other manufacturers pay Lotus to develop their cars and improve the handling, although I believe they're not allowed to mention Lotus's involvement unless they pay a huge licensing fee to use their brand name. You don't hear much about this because it's not in the client's interests to broadcast that information. Lotus built the VX220 for Vauxhall/Opel and the first Tesla – the Roadster – used a modified Elise chassis. I remember testing it on the Lotus track.

It's a unique feeling driving a Lotus and I could probably tell I was in a Lotus even if I was wearing a blindfold in the passenger seat. When the Elise came out in 1996, it was a stunning car. It was really light with a simple but zingy Rover engine so it felt fast, but the handling was something else. They developed it into the Exige, which was the hardtop version, but they've basically been surviving on a single car in a very small niche for such a long time. That's why they've struggled and, a bit like Maserati, they've been passed about through different ownership, which has been a bit of a bumpy ride. That's ironic, given that that's the last thing you feel in a Lotus.

When you're buying cars, the apprehension about whether or not the company's going to go bust does become a factor. If something goes wrong with the car and the company's not on the other end of the phone, what happens then? This problem happened with the Ocean, produced in 2023 by US car company Fisker, founded by Danish automotive engineer Henrik Fisker. The Ocean was an electric SUV built as an alternative to the Tesla, but it wasn't fully developed so it would encounter major software glitches. Unfortunately, for both anyone who worked at Fisker and for anyone who bought the Ocean, the company went bust in 2024. Unlike with an internal combustion engine, where you would be able to find a mechanic to fix the problem, in an electric car, where the energy flow and the battery management is all controlled by software, you're shafted if the car bricks itself.

That means that some of the people who've bought Fisker Oceans are now left with expensive ornaments. And because they're quite heavy cars, they can't even move them off their driveway.

Lotus have been majority-owned by Chinese automotive conglomerate Geely since 2017 and their cars have changed a lot. The Emira, released in 2022, is a pretty car – it looks like a Ferrari – and it's decent to drive, even though it's quite a bit heavier than previous Lotuses, at nearly 1,500kg. The trouble was that after production delays and inflation, the V6 version of the Emira ended up ballooning in price to over eighty grand. That meant that it wasn't rivalling the Porsche Cayman any more, which was the original plan. It was now competing with the 911, and that's a different game. And that's one of the reasons why the Emira has dropped in value an awful lot. And when that happens, it's going to deter even more sales because car buyers are thinking about future values a lot more than they did in the past.

The Emira was claimed to be the last Lotus with an internal combustion engine. The other cars in their range are both dual-motor electric cars. There's the Eletre, a full-size SUV released in 2022, and the Emeya, a five-door grand tourer (their version of the Porsche Taycan, basically) released in 2024. When you look inside them, they're lovely. The design is pretty cool and the infotainment system is really fast because the Chinese can't half do tech. They're fast but they're heavy, heavy cars that don't feel like a proper Lotus because they're a little bit unstable when you're pushing them hard. The reason is that the stability control system – which stops you skidding – hasn't been fully developed. You still notice that a bit with Chinese cars. They've got the battery range, the looks and the interior tech, often better than the Western companies, but in terms of how they feel to drive, they're a bit off the pace. I did a launch test in the Lotus Emeya and as soon as I started experiencing any kind of traction issues, it felt like it was going to spit me off the road. They haven't spent enough money to get those bits right. By comparison, the car that the Emeya is competing against in the market, the Taycan, was rock solid in that situation. That said, there's a lot that I like about Lotus's electric cars and I'd consider one as a company car lease to try something a bit different. As a private buyer, though, it's going to cost you over 100 grand new and then within a year you're done for forty grand. If you want a Lotus, you're better off buying a much older one.

What doesn't quite sit right is that electric cars are the opposite of the ethos Colin Chapman pioneered. He wanted lightness and handling, but Lotus's electric cars are all about power and weight. Geely want to use the Lotus badge because it does carry some cachet around the world and people do recognise it. But apart from the straight-line acceleration, there's nothing in common with previous Lotuses. In 2019, Lotus unveiled an all-electric hypercar called the Evija. I saw one at Topaz Detailing having PPF (paint protection film) applied and it looked amazing. It had the stats to back it up, with its four electric motors dishing out an eye-watering 2,011bhp. Jenson Button had a bespoke one delivered to him in California with detailing that paid tribute to his World Championship-winning Brawn F1 car. The Evija is thought to be the most powerful production car in history, with a £2.4 million cost to match. Amusingly the battery pack alone weighs over 753kg. How much did the Lotus Elise Mk 1 weigh? 725kg. Colin Chapman would be turning (quickly and lightly) in his grave.

Top Gear did a road and track test on the Lotus Evija in April 2025, but curiously the video took months to hit a million views and that's mad for the most powerful production car in history. If it'd been a Bugatti, it would have had loads more views. Part of that's down to the brand prestige and reach, but part of the problem is that it took six years – due in part to Covid and software issues – from the unveiling to the appearance of the first Evija models. It was always going to be a bit underwhelming when it finally arrived.

Lotus PR took me out to China in 2023 to drive the Eletre and we went to the Shanghai Motor Show and to the Lotus factory in Wuhan. The guy who's in charge of the factory, who watches our videos, came up and introduced himself, which was cool. We went to lunch in a nearby restaurant, which was beautiful, although we had to sit over two tables because we couldn't get one table big enough. Never has such a minor inconvenience benefited me more. So I'm not sure what they'd been served, but the people sat around the other table got very suddenly and very violently sick. The poor PR guy – who used to be my boss at *Auto Express* in the noughties – was on that table. He was supposed to take us on the track test and liaise with the Chinese after lunch, but he had to go to one of their meeting rooms in the factory and lie down. I really felt for the guy, seeing as he seemed to be getting a hard time for having the squits. Usually, as a motoring journalist, you're angling to get on the right table, to sit next to the right executive, to try to get the story. Never in my career has being on the right table meant not spending twenty-four hours spewing and crapping through the eye of a needle.

I've had a soft spot for Lotus since that JPS F1 car and because one of the cool kids near me in Walsall had a two-tone Lotus Esprit Turbo with gold writing, which looked awesome. These fond memories of Lotus give them some credit in the bank, in my mind at least. I wonder if the Italians feel the same way about Maserati and Alfa Romeo on account of their proud racing heritage. It's going to be interesting to see what Lotus do next, because the Chinese have a lot of money to invest. They made 12,000 cars in 2024 – that's 70 per cent more than the previous year. So they're doing something right.

Maserati

Full disclosure: I forgot about Maserati in my initial list of cars for this book. And that feels like part of Maserati's problem.

When I was a young chartered accountant, one of my mates I worked with, whose dad owned an accounting practice and had quite a bit of money, had a red Maserati Karif. It was a twin-turbo luxury coupé, produced between 1988 and 1991, and they only made 222 of them. It was one of the fastest cars I'd been in and my friend couldn't drive it at all. But that didn't dissuade me from getting lifts with him just for the thrills of going in the car. Every time I got in, I knew that this was probably my last ride. It wasn't that he was especially fast – he was just unbelievably bad at understanding how to handle a car like that. I almost performed a citizen's arrest on the charge that he wasn't worthy.

In my early days as a motoring journalist, the Maserati that was out at the time was their 3200 GT (made from 1998–2002), which had some sexy boomerang lights, although they changed the look of those, which was a bit of a shame. I remember that car being flawed because the handling wasn't that great and the gearbox was a bit meh, but it had loads of performance and it looked pretty cool. In 2007, they released the GranTurismo, which was styled by Pininfarina and once again looked really cool both outside and inside with the Italian leather interior. It also had an amazing-sounding engine, thanks to Ferrari, who made the V8 for them. That noise was amazing – you could play the throttle like a brass musical instrument. The car was quick, but ultimately Maserati didn't have enough money to make everything about the car great. It's a similar story with Alfa Romeo. One part of the picture always seems to be missing.

The name Maserati makes it sound both expensive and luxurious, doesn't it? Unfortunately, none of their cars have ever quite been as good as I'd hoped. Except one: the fifth-generation Maserati Quattroporte, which came out in 2003. The name 'Maserati' is so good that they can call the model itself something ridiculously boring, like Quattroporte, which means 'four-door'. That's the kind of name Ronseal would come up with.

The Maserati Quattroporte was such a different-looking car to the BMW 7 Series and the Mercedes S-Class and it had more passion and more soul than its German equivalents. It had a naturally aspirated V8 engine and it was a lovely, lovely car. It felt special. I went on the Quattroporte launch in Modena, Italy and we were taken for a wander around a nearby village by the Maserati PR guy. We were just passing this big house and the PR guy stopped us all suddenly and pointed at an upstairs window. 'It's the Maestro,' he said quietly. We looked confused. 'Pavarotti!' he said. We looked up and saw a silhouetted big bloke with a black beard. Now, given that Pavarotti did live and die in Modena, this could be true. Or it could be bullshit that I've been regurgitating to my friends and family for years.

When Maserati tried to do a BMW 5 Series rival, the Ghibli, from 2014, it was a poor attempt. To be fair, though, whenever any company tries to get into that territory, they always fail. Like Jaguar with the XF, which to be fair was a decent car, but not enough people wanted them and they weren't quite good enough. Maserati broke into the luxury SUV market in 2016 with the Levante, which was a bit of a meh car, but they released a sportier model in 2021 called the Trofeo. Some prices stay in my head. That one did, because it was £124,940 and not included in that was the hand-painted Italian flag stripe across the body, which would set you back eight and a half grand. What you do get for that, though, is a 3.8-litre, twin-turbo V8 made by Ferrari, which is basically the one they used in the Portofino. Unlike almost every other Maserati I can remember, there was nothing that stuck out as being especially bad. I was pleasantly surprised because, in all honesty, I did think it was going to disappoint again. It was a step up from the other Maserati SUVs I've driven, where I tend to have the same response each time, namely: 'Love the engine!' But then I seem to run out of gas.

Maserati discontinued both the Levante and Quattroporte range in June 2024 – their last two cars featuring models with V8 engines. The party line was that it's part of their plans to go all-electric by 2028. Carlos Tavares, the CEO of Maserati's parent company, Stellantis, told reporters in June 2024: 'We cannot afford to have brands that do not make money.' Stellantis don't reveal the individual profit/loss figures of any of their brands, but Maserati release their own figures, which they did in July 2024. It turns out that they are facing an €82 million loss not least because they'd only sold 11,800 cars in the first part of 2024 compared with 20,000 the previous year. It was pretty clear who Tavares was pointing his finger at.

Tavares fired Maserati's CEO, Davide Grasso, in early October 2024. And then, at the Paris Motor Show the following week, Tavares broke cover, saying: 'Maserati is in the red. The reason is marketing.' But it won't be Tavares wielding the axe because he suddenly resigned on 1 December 2024 after a big boardroom bust-up. In February 2025, Stellantis revealed that their profits were down 70 per cent on last year, so the rest of 2025 is going to be a big year for them. But Tavares isn't wrong about Maserati. They've got to sort out their identity, pronto.

Mazda

I've really liked Mazda since I bought a first-generation MX-5 in 1997. I had a little bit of money because I was working, and it was a toss-up between a VW Golf VR6 and an MX-5. I was edging towards a convertible, though, because I'd lusted after MGs for years. I'd tried to steer my dad towards buying another convertible – he'd owned an MGB before I was born – and he did test drive a Ford Escort Cabriolet but I think he might have rumbled that I had designs on it.

I test drove the VR6 and it was all right but a bit predictable. I didn't even need to test drive the MX-5. I saw it, knew I'd like it and bought it. And sure enough, I liked it. It had a cool numberplate as well: M447 TEO, which I read as MATTEO. It was a white (no one wanted white cars in the 1990s) 1994 model and came with the basic spec, because that was all I could afford. I bought it from a dodgy bloke in London – it was one of those under-the-arches sort of places where something seemed a bit *off*.

I was twenty-five and made a catalogue of rookie errors buying that car. First, I paid nine grand for it and bought it with a debit card. A credit card would have been more sensible because it would have given me additional rights under the Consumer Credit Act, which makes it easier to get your money back if something goes wrong with the car. I was also a bit scared of the salesman and basically paid the price he plucked out of the sky. I also revealed I'd come down to London on the train from the

Midlands to see this car and was really excited about it. He'd got me over a barrel because he knew how much I wanted to go back home in the car. So I was definitely going to leave with this MX-5 for whatever price he told me, even if it was on fire.

It was a few weeks later, when the odometer clocked up to 50,000 miles, that I found two tiny holes where someone had most likely inserted a pin and pushed the dial back. There's no other reason why there would be holes on the odometer. That made me a little concerned so I gave the car a proper check and noticed that one wheel didn't sit in the wheel arch quite right so it had probably had a bit of a bang. I realised that the salesman had clearly seen me as the ideal person to get rid of this slightly dodgy car. I'd been stitched up.

Fortunately, none of that affected how much I loved that MX-5. Not only was it a convertible with cool pop-up headlamps, it was my first rear-wheel drive car, which was more fun because you could do powerslides like I'd seen on *The Dukes of Hazzard*. Rear-wheel drive gave me that sensation of being pushed along rather than pulled along, which is how a front-wheel drive car feels. Rear-wheel drive was so much cooler, especially when you're accelerating out of a corner. There's a reason why Formula 1 cars are rear-wheel drive, so the rear wheels can do the driving while the front wheels do the steering, so each set of wheels isn't trying to do too much and getting confused. Tyres are like men – they can't multi-task.

My MX-5 drove way better than my XR2, with a sporty edge that I would later find out was quite Porsche-like. It had a unique Mazda-ness about it that made me smile when I was driving it. More than anything else, it was fun. I changed some bits about the car, like fitting an air filter and a performance exhaust. I changed the polyurethane steering wheel to a sporty leather one, switched out the gear knob for a chrome one and put some chrome inserts in the dash to make it feel a bit more fancy.

I remember trying to show off in it once, which backfired spectacularly. I was driving around this roundabout in Southend and I saw this girl in a car next to me, so I thought I'd impress her with a bit of a slide round the roundabout. I have no idea now why I thought that would impress her, but I got it all wrong, went past her and then spun the car all the way round so I was facing the wrong way, stationary, while she came past and slowly shook her head at me. It was a low point.

I had the MX-5 when I was working on my local evening newspaper and for a little bit while I was working on *Auto Express*. I ended up 'selling' it to my ex-girlfriend after we'd split up, and she wrote it off. I wasn't smart enough to take the numberplate off it, so if you've got it, give me a shout.

There are certain cars that come along that no one competes with, like the Porsche 911. Some manufacturers do occasionally try it but they always fail. Similarly, no one competes with Mazda on the two-seater roadster. There's no point. In all of its incarnations, the MX-5 has stayed true to its roots, sticking with the four-cylinder engine. The first-generation MX-5 weighed just 1,002kg and the latest model weighs only 50kg more. The one problem they did have is that the first- and second-generation MX-5s used to rust like crazy but they'd sorted it for later models. Initially, some people derided it as a hairdresser's car, but it gained many fans, particularly among people who were into driving. Lots of motoring journalists owned them and loved them. It had a pleasingly simple layout with a longitudinal engine, rear-wheel drive, a really nice gearbox and double wishbone suspension (allowing each wheel to react independently, giving them optimum contact with the road, increasing cornering performance and making for a smoother ride). In 2021, Porsche made a big thing about putting double wishbone suspension on the front of their top track car at the time, the 911 GT3 (992). The MX-5 had had it as standard since 1989.

Back in 2010, I bought a second-generation Mazda MX-5 – the one that looks a bit like a fish and that no one likes – for two and a half grand. It was part of an experiment to see if, with a £5k budget, I could buy a car that could beat a brand-new Porsche Boxster. The Mazda had about 80,000 miles on the clock but it was in really good condition. One of the great things about the MX-5 is that the engine was originally designed to go in a bigger car – the 323 Turbo 1.8 – so they've got extra cooling and are ripe for a good old tune, basically. I bought a bunch of parts online so we could add a supercharger and some brands supplied other bits, like suspension and brakes. It was the sports model, which had a Torsen limited slip differential (also known as a torque-sensing diff). That's a good thing because in a rear-wheel drive car, if you're coming out of a corner, the inside tyres haven't got as much pressure on the ground so they can spin. What this differential does is to make sure, using clever gears, that the power is sent to where the grip is. It's a bit like when you're rock climbing and you drive off the boot where you can feel that you've got grip.

I got my racing driver colleague to take both the upgraded MX-5 and the brand-new Boxster around a track. In the end, the Mazda couldn't keep up but it wasn't that far behind. The Boxster's got bigger tyres and a more rigid structure so you could properly chuck it into the corners. Still, for a second-hand £5k car versus a brand-new Boxster costing around £50k, it was a good effort. I've still got the car now. I do think about selling it but it'd only fetch three or four grand and it's still so much fun to drive. It's such a unique, joyful little thing.

I did have a decent relationship with Mazda, back when I was at *Auto Express*. I liked their cars and they liked me writing nice things about their cars. That positive relationship continued for the first couple of years I was at Carwow. In 2019, we were just starting to hit our stride with over a million subscribers, we were confident and experimenting by being a bit more controversial. Mazda lent me the Mazda 3 to review and, for some reason, when I was looking at the rear of the car, I was reminded of what Jeremy Clarkson said about the back of the Chrysler Crossfire looking like a dog having a poo. And then it came to me: the Mazda 3 did genuinely look like a cat taking a shit. We found a picture of a cat – the same colour as the car – straining on a litter tray, added it to the video and uploaded it. Mazda were bloody furious. They asked Carwow to take the video down. We didn't.

The Mazda 3 video generated three million views and we put it out on our separate Japanese channel and that video did 1.5 million views. The big bosses in Japan would have seen it and given the Mazda UK guys a bunch of grief. The thing is, after that initial silly comment in the video, I actually gave the car a good review because I really liked it. But if someone tells you you've got a really ugly face, before they go on to tell you how much they like you, we all know the bit you're going to remember.

Mazda haven't lent me a car since. I think I'm dead to them.

In those early days, we were having a lot of fun while we were filming because being at Carwow was liberating compared to working for *Auto Express*, which was more serious. On Carwow, the idea was to talk about cars like you would do with a mate. And this more relaxed style clearly resonated with people.

I didn't need to say what I did about the Mazda 3 – I was just saying what popped into my head like I would when chatting to my car buddies. It's a shame for Mazda in a way because I do like their cars, and I'd be mindful of saying anything as extreme as that again. Mazda videos get decent views because they've got a culty, loyal fan base, including me! Here's hoping they get back in touch at some point.

Mercedes

When your friend asks you to pop round to his lock-up and then puts a blindfold on you at the door, you do wonder if you're going to end up dead in a ditch. But seeing as it was Yianni (my friend and car customiser extraordinaire), it was always going to involve messing about with a car. He wanted me to guess what kind of car was in his lock-up from the noises that parts of it made. So Yianni opened the door – it sounded like a car from the late 1980s or early nineties with one of those fancy handles that pulls out a bit further than the regular ones that just swing upwards. And when he shut it, I just thought that sounds like a *German* shut. I figured it was a saloon boot because it sounded heavy but not enough to be a hatchback. I was thinking Mercedes.

Yianni started it up and I could hear the four beeps signifying a crappy old immobiliser. Next up was the spoiler, which felt foam-like and squidgy. Definitely a sporty Mercedes from the late eighties or early nineties – probably 2.5-litre and, knowing Yianni, I bet it was the Cosworth variant. I could see the car in my head. Obviously, you doubt everything you've ever learned once the 'reveal' takes place, but as the roller doors opened painfully slowly, there it was: the 2.5-litre Mercedes 190E Cosworth.

It turns out that I have a very particular set of skills and I will identify your car.

That 190E Cosworth was a fast car. In my late teens and early twenties, I was really into motorbikes and bought myself a Honda VFR400 NC 30, which could do 0–60 in under four seconds and reach 140mph. I remember it had a very 'tall' first gear, so you could clear 70mph, which was ridiculous. We're talking supercar-at-the-time performance. So you can imagine my surprise when I struggled to shake a Mercedes saloon racing around the Walsall ring road, although I did have my mate Greg on the back of the bike, which kind of killed my power-to-weight advantage. But not being able to outrun a supposed grandad car did made me think there was something wrong with my bike. Greg explained that this version of the Mercedes 190E had a Cosworth engine, so if any grandad did own one, we were talking about someone of Sir Stirling Moss's calibre. Since then, the 190E Cosworth stuck in my mind.

It's not the Mercedes that comes into my mind first, though. That would be the 300 SL, which I love and, to be fair, who doesn't – it was the first production car to feature gullwing doors. The 300 SL started life as the W194, which won 24 Hours of Le Mans in 1952. The chassis was based on Anglo-German engineer Rudolf Uhlenhaut's pioneering space-frame design from 1947, which saved a lot of weight. In a reverse of the 'you're only supposed to blow the bloody doors off' line from *The Italian Job*, the Mercedes engineers were struggling to put the bloody doors on because there were two big bulkheads along the side of the frame. It was at this point that someone slammed down the seventy-three-page rule book on the drawing board (they would have done in the movie, anyway). The relevant bit read: 'At least two permanent doors had to be fitted on either side of the body in such a way that permitted proper and direct access to the front seats.' Nothing about how the doors had to be connected to the chassis. Cue light-bulb moment. So they hinged the doors from the roof, and it became a legendary design. Although, having climbed into a 300 SL, I can tell you that 'proper and direct access' is a bit of a stretch. But once you're in, you don't want to leave.

Moving into the 1980s, the iconic Mercedes model for me was

the red 450 SL (R107) – the one Bobby Ewing drove in *Dallas*. I loved that car. Design-wise to me, Mercedes at that time were just on another level to any other car brand with that quite squared-off Bauhaus look, which I've always admired.

In the early 2000s, when I started as a motoring journalist, Mercedes were using their amazing 5.5-litre supercharged V8 engine on models like the AMG SL 55. I wasn't allowed to drive that car because I was too junior at the time, but I was begging to. The next best thing was getting one of my senior colleagues to give me a go in the car so I could listen to that crazy supercharged engine. It reminded me of the sound of a low-flying Second World War German fighter plane. And it turned out that wasn't too far off the mark, given that the most famous German fighter – the Messerschmitt Bf 109 – featured a supercharged Daimler-Benz engine. Engine quality is something that had remained consistently high with Mercedes, especially with their diesels. Mercedes were the first car manufacturers to build a production car with a diesel engine, the 260 D, which was the star of the Berlin 1936 motor show. That model became the taxi of choice in Germany, partly thanks to its incredible fuel economy. Mercedes also developed the first turbo diesel production model – the 3.0-litre, five-cylinder 300 SD from 1978. So they have some track record.

There are lots of reasons I really like Mercedes. But I noticed that during the early part of my career as a motoring journalist, the quality of Mercedes started to go downhill. They weren't quite so over-engineered any more. Before then, Mercedes cars felt like they were hewn from granite but they were losing that slick finish that differentiated them from the competition. Then, around 2015, they came back strongly with new interior designs, which had a cool, avant-garde feel to them.

More recently, Mercedes adopted an all-in approach on EVs, betting heavily on rising demand and setting up two different production lines, one turning out internal combustion engine cars, and the other making electric cars that were bespoke from the ground up. It was a pioneering move for one of the legacy car manufacturers and, at the time, I thought they were doing the right thing. It seemed like they'd opted for the 'purer' approach, which would allow them to make the better EV while, by contrast, BMW's shared-platform strategy felt like a compromise. But in the end, having the two production lines meant that the cars were expensive to produce so Mercedes priced them too high.

Yeah, Mercedes got burnt by the 'meh' uptake for EVs among the motoring public, but they also didn't get the design right. It felt like Mercedes spent all their time focusing on efficiency and reducing drag, while looks didn't get much of a look in. The upshot was that their EQ range looks like someone left them too near a radiator. Maybe Salvador Dalí would have been impressed, but no one else seemed to be. Mercedes cars are supposed to be imposing, elegant, executive cars. But the EQ range is just a series of anonymous shapes. A big part of the appeal when you buy a car is what it looks like, and when you're paying top-dollar for one of the EQ range, they've got to look the part.

Mercedes have been all about cutting-edge technology and luxury, but they're being left behind on electric car tech. That said, they are catching up with their new CLA, which has a claimed range of 500 miles on a single charge. For a long time, though, Teslas were way more efficient than Mercedes' EV offerings. I remember speaking to a Mercedes engineer about exactly this. 'Your EV models are loads heavier than Teslas. What's going on?' I asked. He explained to me that part of the problem is that they have to use the same control modules and wiring in their EV cars that they use in their petrol and diesel cars. So even though their EVs are built from the ground up, they're making a compromise when it comes to the whole vehicle's electrical system and this means more modules and more wiring, which adds weight. Mercedes, like other mainstream European car makers, effectively have a blind spot because of their legacy. And we can turn the blind spot metaphor into a real-life example because if you were to design the human eye from scratch, you wouldn't have the optic nerve coming out of the middle of the retina, which is the reason we have a blind spot in the first place; instead, you'd configure the optic nerve so it enters from the side and then you wouldn't have a blind spot at all. The Chinese firms and Tesla don't have a blind spot because they start their design process from scratch. Mercedes have been behind the curve on tech and yet they're still charging premium prices. It will be interesting to see how things change with the new CLA.

When I was growing up, we didn't see many Mercedes. Most of our friends and family had Fords and Vauxhalls. There was one Mercedes on our street, tucked away in the driveway of the poshest house. It was a 280 SL (the R107) and that was such a beautiful car. It's still such a beautiful car today. Mercedes weren't really a brand that you were dreaming of as a teenager, though. BMW were focused more on the thrill of driving, but Mercedes pitched themselves as a luxury car for a smart, older gentleman. It changed when Daimler-Benz extended their racing-car collaboration with AMG into production cars. That started with the C 36 AMG in 1993. Those AMG models were such a hit that DaimlerChrysler (as the Mercedes-Benz group was known then) bought 51 per cent of AMG in 1999 and the company became Mercedes–AMG.

I've had quite a few Mercedes as long-term demo cars, including the Mercedes-AMG G 63 G-Wagon. One of the rare benefits of Covid was that the six months that you usually get with a long-term demo car (you can't have them any longer than that for compliance reasons) went out the window, so I ended up with that car for a whole year. It was the perfect car for Covid because if we were facing the apocalypse, at least I had a G-Wagon to smash through all the zombies. It was an amazing car with an amazing engine – the 4.0-litre twin-turbo V8 – and pretty much any Mercedes that engine went in was a win for me. I love the AMG attitude, which is a distinct personality, compared to a BMW M car or an Audi RS. They're everything that a petrol head likes about a car: they're in-your-face, lairy, hairy-chested and bonkers. And Mercedes can be absolutely bonkers.

One memorable example was Mercedes allowing me access to their heritage fleet over in Germany. They were happy for me to do 0–60 mph launches on some of their classic cars from the past, like the iconic *Grosser* (the 600 – the forerunner of the Maybach marque, which back in 1963 was the priciest car in the world) and the CLK DTM, inspired by the touring car that won the 2003 DTM season. It isn't the first time Mercedes have done something crazy. They're famous for it.

In the late nineties, Mercedes were just about to launch the Maybach 57 and the longer Maybach 62, two luxury limousines built to compete with Rolls-Royce and Bentley. The legend goes that when the Mercedes board were deliberating about whether or not to sign the car off, one of their assistants came running into the boardroom. The A-class, which some of the execs were worried was going to downgrade the Mercedes brand, had just rolled in the elk test (an evasive manoeuvre test involving swerving to avoid an obstacle). Nightmare. To offset the negative publicity, they decided they needed to do something extravagant to prop up the other end of the brand, so they signed off the Maybach and went all-out for the launch. The invited journalists were flown out to New York on Concorde and then invited to board the *QE2*, which travelled back to the UK with a Maybach 62 in a glass case on the deck of the ship. Guttingly, having only been at Auto Express for a year at that point, I didn't make the cut. I'm not saying I'm praying that another Mercedes spectacularly fails the elk test, but it could work out nicely for me now.

At first, some manufacturers weren't interested in Carwow because we weren't big and were looked down upon, but Mercedes were very supportive. It seemed like they had a good sense of the opportunity that social media presented and always lent me press cars, when certain manufacturers refused to. I remember that one of the first Carwow videos that really took off was a review of an E-Class.

In some ways, I've been a bit unlucky with Mercedes cars. It started when I talked Mercedes into lending Carwow a bunch of high-performance models to do a big drag race. We filmed it and it went really well but we discovered loads of stones on the runway (one we'd never used before) too late and, well, we destroyed half the cars' windscreens. When you return cars to a car manufacturer's press office in that state, not only do you have to pay for the damage – you've also screwed up the arrangement for the next person loaning the cars because everything's always on a tight turnaround.

It takes a while to regain a manufacturer's trust after a balls-up like that. And we'd just about managed it, so we asked Mercedes to lend one of my members of staff a brand-new AMG GT 63 four-door to review, which they did. As part of the testing, he took the car on a route to his house which was on a private road, and managed to slide off the road and straight into a tree, causing forty grand's worth of damage. And because our insurance wouldn't cover an accident on a private road, we had to pay for it. It sounds like these accidents happen all the time. They really don't. They just always seem to bloody happen when I or another member of my team is filming with a Mercedes.

Sometimes it is our fault, though. When you drag-race cars,

manufacturers like to know in advance which cars they're going up against. They don't want their car being set up as some kind of sacrificial lamb or used as the butt of a joke. On this occasion, Mercedes had lent us an AMG GT 63 S and we raced it against a Porsche, which they knew about. But after we'd done that, we had a bit of extra time, so we decided to drag-race it against a Hyundai Ioniq 5 N that we were using in a separate race. We imagined it would be very close and make for a great video. As it turned out, it was – the Ioniq 5 N edged the standing quarter-mile by millimetres.

Mercedes will never stop you doing what you want with a car that a customer owns, but the problem was doing it with *their* car. Ultimately, they've got certain rules in-house about the message they want to broadcast, and that won't include putting up a new AMG GT 63 S against a £60,000 electric Korean hatchback. They don't want to make any association with that car. It was our fault for forgetting to tell Mercedes. Normally we'd ask permission from a manufacturer before doing this, but in the heat of battle on the day, we forgot. The extra dimension was that we'd also used a guest driver, which caused all sorts of problems with compliance law because he wasn't on their insurance forms. We didn't realise that we needed to supply that information, because we ordinarily insure people at the track, but it turns out that we did. After that, we weren't given any Mercedes cars to drag-race for a while. To be fair, they did still lend Carwow cars to review and invited us on launches, so they didn't completely cut us off.

In this relationship I've been a shit boyfriend to Mercedes. I haven't treated them right. I apologised and told them I was going to change my ways and they forgave me. And then I went out late with my mates again. It's been about a year since the incident and finally Mercedes have forgiven me. They will once again lend us cars to drag-race. Hopefully I won't cock it up again.

If I could own one Mercedes, it would be either the V8 560 SL (R107) from the late 1980s or the SL 63 AMG Black Series with the gullwing doors. The SL 63 AMG has a naturally aspirated V8, it sounds incredible, it has a brilliant throttle response and is an amazingly beautiful car. What they did to the Black Series is more than the sum of its parts. The changes they made over the standard model, like transforming the handling, was well worth it. It's a brilliant car.

I do want to give a special mention to the AMG ONE because Mercedes went mad with this car and I got to drive one somewhere special. The idea behind the AMG ONE was: let's put an F1 engine in a road car. What could go wrong? Well, an F1 car has a team of people operating it. A road car version has you. It does sound problematic. And then you've got fiddly road-legal details to worry about, like passing an emissions test and going over a speed bump. All of this shouldn't be possible. And yet, somehow, they pulled it out of the bag.

I got an email from the guys at Mercedes inviting me to test the AMG ONE. Amazing – I knew they'd only be inviting maybe three or four UK journalists. But then I saw where they wanted me to come to test it: the Nürburgring. I've driven the Nürburgring before. Jackie Stewart nicknamed it the 'Green Hell' for a reason. Built in the 1920s, it's the world's most dangerous racing circuit, thirteen miles long, with 300 metres of elevation from its lowest to highest point. Everyone crashes there. But not everyone crashes in a brand-new £3 million hypercar. I didn't just have that to worry about, though. I'd also be trying to get to grips with the car, assess it and actually shoot a good video, all the while praying that it didn't rain, because when it does, that track turns to glass.

I got there, with the fixed smile of a man who's crapping himself inside, only to find out that it wasn't the Nürburgring we'd be testing on. It was the Nürburgring GP circuit – the three-mile track constructed in 2001 entirely from asphalt. I hadn't read the invitation properly. Cue a massive sigh of relief, followed by a realisation that I didn't know this track at all and was about to test a £3 million car on it. Usually, if you're testing a car on a track, you'll have studied the track beforehand and prepare for the turns and trickier bits. But here, I was trying to learn the track on the fly while presenting to camera and reacting to what the car's doing. And it's not a great start when you spend five minutes trying to turn it on before a Mercedes engineer feels duty-bound to tap on the window.

So when I eventually pull away, the car's being powered by the two electric motors, each one giving you 163hp. But then the F1 engine kicks in, like a pneumatic drill behind your head (but in a good way). Yes, it's fast, but I wasn't expecting it to be so easy to drive and to give you, the driver, so much confidence. But it's too complex for its own good. On the launch, there were two cars, and Chris Harris, who was filming for *Top Gear* TV, was in the second one. That car had no end of problems – to the extent that they struggled to make a video. I was luckier, but my car had a few issues too. At one point, the engine just cut out. One of the engineers told me that in the mode I was in, the battery was being continuously charged by the engine, so to stop it from over-charging sometimes the engine cut out. I think they must have been having issues with the energy control system because this shouldn't happen. The cars were clearly so complex that even on the launch they were having teething problems. In fact, many supercar dealers I've spoken to have told me there are often issues with AMG ONEs.

That said, the AMG ONE is an amazing machine to drive. Vents pop up on the bonnet, a huge wing appears on the tail and the whole thing lowers by 30cm. It's some piece of engineering.

If money was endless, I would like an AMG ONE. I mean, it's got the same engine that Nico Rosberg won his F1 championship with. In a road car. It spits out 1,006hp. Everything else is meaningless. Except that, true to form, on a trip to Dubai where a benevolent billionaire lent us a whole bunch of supercars to drag-race, including the AMG ONE, guess which one broke? And it wasn't me this time. The AMG ONE has got problems, but what do you expect from a car that is utter madness. It should have an entire team of F1 engineers supplied with it but instead, it's being driven by people who can't operate a spanner. It's a crazy, crazy idea. But Mercedes are brilliant for doing it.

MG

My dad bought an MGB roadster in the mid-1960s. The one with a chrome bumper. It was red, but he had it resprayed orange. He sold it after I was born and it was the one car he had that I wished he'd kept. Never mind that we wouldn't have all fitted into it. The real shitter was that my dad's mate, who bought it off him, wrote it off twenty-four hours later. I've been mad at this bloke ever since, although I never met him. I felt like I'd been denied my birthright so I grew up wanting an MGB. So, around seventeen, every mind-numbing hour that I worked at Argos in Walsall on minimum wage was all going into the MGB kitty. I went to look at quite a few and almost bought one, but my dad, of all people, put me off. 'It'll be loads of trouble,' he told me.

Somewhere along the way, I ended up preferring the smaller MG Midget to the MGB because I like the compactness, the dashboard and the little farty-sounding exhaust. The problem is I'd need to build a garage because if I kept it outside, it would suffer the fate of many an old MG: it would dissolve. You can underseal them but they'll find a way to rot away. And once they start, you're entering a world of pain: bodywork repairs. I will own one at some point but I'll need to be careful it doesn't end up like my Fiat 126, under a cover for most of its life. That 126 is a fun, silly little thing, but there's the constant risk that it will suddenly cut out. Unlike my Mazda MX-5, which has performance and never misses a beat. The MX-5 is quite like a MG. An MG that works, that is.

MG were part of British Leyland in the 1970s and 1980s, which became the Rover Group in 1986. The MG badge was basically used for sportier versions of Austin Rover models like the Metro and Montego. When I was at school, we had a Metro, but god I lusted after the MG Metro Turbo. It had red seat belts and was faster than the regular Metro. There are certain points in your life where the want hurts you, and this was the second time I'd felt that. The first was the burning desire for a Raleigh Ultra Burner bike. The silver and blue one. I wasn't thinking *I really want a 911 or a Ferrari* at that point. I kept my horizons low and faintly achievable. And the fact that the MG Metro Turbo had about 90hp didn't matter. Compared to the Metro we had, that one was epic. Even now, I'm tempted to get one. But there's another, louder voice in the back of my head, saying: *I bet it's shit.*

After BMW bought the Rover Group in 1994, the MG marque passed to them. The year after, the first MG since the MGB rolled out of the factory, the MGF. I thought seriously about buying one of these at the time. It came down to the MGF or the Mazda MX-5, and I wanted the MGF more because its variable valve timing made it a bit faster than the Mazda. In the end, I went for the Mazda because I knew it would be the better car (and having driven both, back-to-back, it was the better car by far). I'd started to think not just about what a car looked like, but what it would actually be like to own. I did my research and knew that the Rover engine in the MGF would suffer head gasket issues and the engine was mid-mounted so any problems would be a nightmare to get at. Plus, the Mazda had the pop-up lights, and pop-up lights always win. So when I was finally in a position to buy an MG, I bought a Mazda.

In the noughties, MG did some crazy stuff when it was part of the MG Rover group. The MG XPower SV was one such project. MG Rover had swallowed up the Italian car company Qvale, who made a front mid-engine, rear-wheel drive car called the Mangusta, and then used that as the platform for a car that was utter folly. The XPower SV was manufactured in Modena and finished at Longbridge, which doesn't sound dissimilar from 'New York Paris Peckham' painted on the side of the Trotters' yellow van. The MG XPower SV had a 5.0-litre V8 engine but it was a parts bin of a car, harvesting bits from the second-generation Fiat Punto, MG TF and Rover 75. They only made about eighty of them, including the upgraded SV-R model, on sale for about £83k, one of which was bought in pearlescent racing green by petrolhead Rowan Atkinson. Because they're so rare now, a good one would sell for around £130k in 2025.

MG Rover went bust in 2005 and the key assets were bought by state-owned Chinese firm Nanjing. Two years later, Nanjing became part of the largest Chinese state-owned automobile man-ufacturer, SAIC. That year, when I was still working for *Auto Express*, I went to the MG relaunch at Longbridge. They had a couple of MG TFs there (the updated version of the MGF) and I remember being told that MG (which actually stands for Morris Garages) now stood for 'modern gentleman', which made me feel a bit sick. Not a great start. Anyway, it was mostly UK media at the event, with some European and Chinese press there as well. I remember nipping to the toilet and coming back to the room where I'd just been, but the UK media had all been very quickly taken somewhere else and it was just Chinese media there. It all seemed a bit odd to me, but then again, our two countries do have slightly different attitudes towards press freedom.

The latest MGs are decent. The initial petrol offerings weren't so good, but their electric range gave pretty good value for money. I was really surprised when I first drove the MG4 EV hatchback in 2023. It's a rear-wheel drive car and it felt fun and quite sporty. It looked pretty decent too. It was less expensive than the equivalent VW, but it was the superior car. The infotainment system was a bit better, but the way it drove was noticeably better. Interestingly, though, the *Auto Express* 'Driver Power' survey, which gives you an idea of what a car's actually like to live with, revealed that MG gets rated pretty poorly for customer satisfaction. From a motoring journalist's point of view, when you're just driving the car in isolation, it all feels pretty good and very well priced, but MG owners do complain about reliability issues. They're true to MG form, in fairness.

Also in 2023, MG started producing the Cyberster, the first electric convertible sports car. I really like the idea of this car, but it left me a little bit cold. It's got good performance, but it just doesn't handle as well as it looks like it should. In my review at the time I said it was comfortable and relaxing to drive, but that's probably a bit kind. It's got crazy doors, like a Lamborghini, which actually turned out to be a bit of a faff to operate.

I was filming at Goodwood for the Festival of Speed and they had a few MG Cybersters there, so I took one home to Oxfordshire to review. I was knackered after filming all day, and the last thing you want after a day's filming is an electric car, especially when it hasn't got a full battery. Now, in theory, I should easily have been able to get home. I took the roof down, briefly stopped at a restaurant on the way because I hadn't eaten all day and was back on the road. I was driving on an idyllic July evening with the roof down in a 500hp sports car, but then the range anxiety struck. I realised that I wasn't going to get home without charging it. So I faffed around trying to find a place to charge it but it was getting late and the chargers weren't working. I found another place but it was in a park, which was locked. Eventually I found somewhere to charge it but by then I'd been so consumed by range anxiety that anything that was slightly more difficult than it needed to be was magnified. Like the doors, which were more difficult to open than normal opening doors, and that was enough to make me want to beat the car like Basil Fawlty.

Range anxiety is something everyone who's had an electric car will identify with. My mum had a Tesla Model 3 for three years and loved it but she got rid of it because she never charged it away from home in that time. She'd come to my house, and – bearing in mind that she could get to my house and back to hers without charging it – the first thing she did was plug it in. Before she'd said hello. Range anxiety does funny things to people.

After I'd forgiven the Cyberster, I took it to our local village fete, which I'd been asked to open. I could have taken my Porsche 911 GT3 RS but I figured that because the Cyberster was quite an unusual-looking car, it would get a lot of interest. I parked it in the middle of the village green with the doors up and loads of people went and had a look. 'Ooh, it's an MG. Wasn't expecting that,' was the common response. 'Would you like one?' I asked. 'No,' was, again, the common response.

In 2024, I tried out the MG4 X Power, the performance version of the MG4. Its 435hp tops the Audi RS 3 and the Mercedes-AMG A45 S. It was Porsche quick, clocking 0–60 in 3.77 seconds. For the money you're paying – around £30k (on Carwow anyway) – this car is insane. If you're after a straight-line performance hot hatch for a stonking bargain, it's a no-brainer. The MG4 X Power is a great example of what the Chinese are good at. Unlike European car manufacturers, who have to faff around with two production lines, the Chinese make electric cars from the ground up. They're much more affordable, decent quality, and they're selling well. In 2024, the MG4 was the second bestselling EV in the UK among private buyers.

Bizarrely, though, the high-performance version of the MG4 is less fun to drive than the original and I honestly don't know how they've managed that. It's got the outright thrill of that mad acceleration, but it doesn't seem to handle it as joyfully as the normal car, which is slightly lighter. The MG4 X Power is four-wheel drive, not rear-wheel drive, and has firmer suspension and so doesn't go down the road quite so well. In fairness, I wasn't expecting the original MG4 to be as good as it was and so my expectations had increased when I stepped into the high-performance version. But in the MG4 X Power, after you've got bored of the straight-line performance, you find yourself underwhelmed. I preferred the standard car and I can't remember the last time that happened. It's certainly not something you'd say about the BMW M3.

Yes, the latest MG offerings have pleasantly surprised me. Do I personally want an electric MG? No, but when I'm doing consumer reviews for people who are looking for an affordable, good-to-drive, decent-value EV, I'd say yes, get an MG, but check the customer satisfaction surveys first. I would say that in the UK, MG have been the best electric car brand around, for the simple reason that they came in and did it cheaply and well, which is all that people wanted. Because frankly every other electric car is just bloody expensive. Until recently, that is. The new Renault 5 E-Tech looks cool, is really good fun to drive and it's on sale for a decent price. So, MG may need to check their door mirror because the Europeans may – finally – be catching up.

Mini

As long as I've been alive, Minis have been cool. The first car I remember my mum owning was a Mini Clubman, which had a squarer shape and a more protruding bonnet than previous Minis. I loved the shape of the Clubman – it just looked a bit more badass than a regular Mini. But Christ, when she bought that car, I'd never seen something in such atrocious condition inside that hadn't been in a crime drama. To be fair, it wasn't that much better when we cleaned it up, with its beige vinyl that used to burn the backs of your knees in summer. That was an era when anything approaching what we'd now call health and safety was mocked. Wearing a seat belt? Wimp. Wearing a bike helmet? Loser. The prospect of serious injury didn't really enter anyone's head, so I had the time of my life bouncing around the back of that Mini like a dog on a trampoline.

My rich mate's mum had the 1275 GT, the replacement for the Mini Cooper, with the funky graphics on the side. If the Clubman was cool, the 1275 GT was sub-zero. People forget that Minis were very clever cars for their time. Sir Alec Issigonis, who designed it, was a genius. He didn't create the first transverse front-wheel drive engine (as is sometimes claimed), but he did position the transmission directly underneath the engine, and they shared the same oil reservoir. It was a space-saving masterstroke, allowing the car to be only 4 feet wide. It's incredible how small a Mini actually is, and it's something you only realise when you squeeze into the back.

When I was old enough to drive, I started in my dad's Mini Metro. It was depressingly slow and I remember trying to pull out the choke to make it go faster, not really understanding how a choke worked. The Mini Metro made a very distinctive whining sound when it was in reverse, so I could hear that my dad had arrived to pick me up from my friend's house. So for many years, a Mini Metro in reverse screamed 'Fun's over!' There are some cars that make unique sounds. You can hear a Subaru in a different county.

Loads of people had Minis as their first cars. Yes, they had loads of problems with the electrics and rust and god help you if you were involved in a collision, but they were cheap to fix and such fun to drive. That fun feeling was one of the key aspects of Mini that BMW retained when they developed the new Mini, which they launched in 2001. But where I think the Germans really smashed it was with the marketing. They really leant into Britishness and linked the DNA of the new Mini with the iconic Minis from the 1960s. It acknowledged and respected the past but looked to the future.

I'd just started at *Auto Express* when we awarded the new Mini the Car of the Year gong in 2001. That car managed something that very few small cars have achieved: rich people wanted them too. And that was because it didn't sacrifice desirability in favour of economy. It was a smart, stylish, executive supermini that was such fun to drive. I had the Cooper S convertible and it was a brilliant car. The high-pitched noise the 1.6-litre supercharged engine made was something else. And it felt like a Mark 1 Mini from the 1960s with its pointy, go-karty feel, but its suspension was grown-up and refined. It was a trip down memory lane with all the comforts and sophistication of the twenty-first century.

The retro-modern styling, especially the oversized, centrally mounted speedo, was such a good touch. It made middling hatchbacks like Fiestas and Polos look ordinary with its high-spec BMW interior. You'd happily take a Fiesta or a Polo, but you *wanted* the Mini. The new Mini also kicked off the whole personalisation thing with stripes, different wheels, paint schemes and roof colours. It made people happily part with more than the starting price, or what we call in the trade 'price-walking'. But people weren't unhappy about it because those trimmings were desirable and you just wanted them in your life. Mini had created a car that you bought with your heart not your head.

The only fault with the new Mini was something no one could have predicted or prepared for. It was too good for its own good. It was like they'd scaled a summit by creating a modern classic and whichever step they took next was downwards. As we all know with iPhones, you've got to introduce desirable changes every couple of years so people upgrade to the even cooler next-gen model. Most car manufacturers do this by announcing a facelift after a few years that's always been part of the original sales plan. But if you throw them everything you've got at the beginning, sales will be terrific for a while, but there's nothing left in the locker. With the new Mini, sure, there were minor improvements to the fit and the finish, but new crash regulations lifted the body up and made the shape a little awkward. Then Mini started adding extra bling to them, but it wasn't elegant. It was like they'd ram-raided Claire's Accessories and left with whatever was stuck to the car. In the meantime, other manufacturers were catching up in the cool stakes. When the Fiat 500 came out in 2007, they sold the whole of the first year's cars in three weeks.

At that point, I was working at *Auto Express* and we'd just landed a major scoop. Somehow or other the magazine got hold of some details for the new Mini Clubman R55, featuring the rear 'clubdoor', before it was released officially by the manufacturer. Our editor had cut his teeth as an investigative journalist on national tabloids, so his philosophy was that if there was a story, we would always break it.

The clubdoor was designed on the right-hand side of the car because in Germany that means it's on the passenger's side, and so that's the side that would be next to the kerb. They didn't alter it for left-hand-drive markets, which meant that in the UK (where Mini had its biggest market), your kids were getting into the car on the road, which isn't ideal. But I had an R55 Clubman and I really liked it, although I didn't have a child at that point. You can tell it's a BMW product because everyone uses its model designation (R55) rather than using the more long-winded first-generation, second-generation etc. It's more efficient and makes you feel like you're part of a club. Under BMW, Minis became smart, sporty cars that were designed to be fun to drive. They became a sort of front-wheel drive BMW, and that broadened their reach to people who loved the thrill of driving. They appealed to petrolheads.

Confusingly, the Clubman was the successor to the original Countryman – the two-door estate version of the Mini with the wooden trim on the outside that made it look like a Morris Minor Traveller. As a kid, you couldn't help but think, *A wooden car – how cheap are you?!* Although it did look like an unfinished toilet cubicle, the wood wasn't actually structural, and the design's grown on me.

Mini released the new Countryman in 2010, and in the eyes of some motoring journalists it was a bit anti-Mini on account of the size of it. The whole point of a Mini is that it's dinky, fun and cleverly packaged. While the Countryman did supply the Mini quirkiness and all the add-ons you wanted, it felt bigger on the outside than on the inside, plus, after you started adding options, it was a lot more expensive than competitors like the Golf.

The Countryman was one of those examples that brought to light the disparity between motoring journalists and the buying public. While the Countryman was a sales hit, outselling the Clubman two to one and becoming their second most popular car, it wasn't so popular among the majority of motoring journalists. BMW closed the doors on the Clubman in 2024. It had sold over a half a million but hadn't been a big success partly because it wasn't what people were after. What they wanted was a crossover SUV.

When BMW took over Rover in 1994 and acquired Mini, the public reaction was initially negative. *The Germans are buying up our car industry* was the understandable grumble. As it turned out, they've done us a favour. BMW were so good to the Rover workforce, offering generous severance packages to the people they let go. They've been great custodians of brands that mean a lot to Brits, have taken significant risks along the way, but have had the resources, ideas, talent and confidence to take heritage brands back into the black. BMW also did it with Rolls-Royce with the Phantom. And VW achieved the same with the Bentley Continental GT. Each time, the stakes could not have been higher – the future of the brand was in their hands. And they nailed it.

Mini's marketing in the BMW era was on point. They're very PR savvy, pushing the brand and the image at cool events, festivals and launches. Yeah, they spent a lot of money on the branding, but they absolutely got their money's worth. I was always impressed with the Mini PR department because they've got more to deal with than most car brands. They don't just handle the car press – there's also a lifestyle press component because Mini's association with fashion means that they feature in magazines like *GQ* and *Vogue*. They're the custodians of a proud British icon so they have that cultural weight on their shoulders. The Mini plant in Oxford isn't just a factory, it's a historic site and part of the community, so the PR department has all sorts of local politics and union issues to grapple with as well. It's a brand that represents many things to many people.

I can't finish a chapter on Mini without talking about its two most famous appearances in TV and film. Mr Bean's Mini is so much more than a car. It's the perfect reflection of his personality and becomes a big character in its own right. In the first ever episode, Mr Bean gets irritated by the Reliant Robin ahead of him (which everyone could relate to) and pulls off a late overtake, sending the Reliant Robin off the road and on to its side. It's the moment you fall in love with Mr Bean and his Mini. *The Italian Job* would not be the film it is without the red, white and blue minis. Amazingly, every car they used in the film was driven over from the UK. No transporters, no trains, just an epic convoy on a road trip. I remember hearing that when the driver of the white Mini, Barry Cox, arrived back in the UK, he was stopped for speeding by a local copper. The policeman discovered that the car registration was fake, as was the tax disc, and when he opened the boot, it was full of what looked like gold bars. When he tried to explain that he'd just driven all the way back from Turin shooting a film about a robbery with Michael Caine, he must have sounded like a lunatic, so poor Barry spent the night in jail. He was the only person banged up for pulling off *The Italian Job.*

Mitsubishi

OK, straight into it. Yes, the Boxer engine of the Subaru Impreza had a bit more character and noise, but the four-wheel-drive system in the Evo had this amazing yaw control system and the forward transition was a little bit better. Plus, it felt technologically superior and I thought they were much better with their design, so like for like, I thought the Evo was the better car. Also, unlike Subaru, who had this one great design and then continued to destroy it, Mitsubishi's designs were better and more consistent. I had (and still have) a soft spot for the Mitsubishi Lancer Evo and GSR – the models made in honour of Tommi Mäkinen's ridiculous run of World Rally Championship victories between 1996 and 1999. They are such good cars but very expensive now, and I get a kick from those brilliant rally-bred saloons from the late nineties and early noughties with the Toyota GR Yaris. That's why the GR Yaris is such an amazing car. You get all the same feelings as a Mitsubishi Evo in a new car package.

The Colt Car Company (also known as Mitsubishi Motors UK) imported and distributed Mitsubishis in the UK and they'd arrange to tune certain models for the UK market. The FQ-400 was a stand-out one, tuned up to 405hp with a top speed of 175mph. It'd do 0–60 in 3.5 seconds and a quarter-mile in 12.1 seconds, living up to its 'f***ing quick' abbreviation. The Mitsubishi press office have not officially confirmed that's what it stands for, but they don't need to. It does.

I had quite a close relationship with Mitsubishi and they'd let me borrow their Mitsubishi Evo press cars quite a few times, and even their Tommi Mäkinen Evo VI, which I drag-raced just before they sold it. If ever there was an antidote to a crap day at work, it's spanking an Evo VI. What a car. Unfortunately, Mitsubishi also managed to produce one of the worst cars I've ever driven – the Mirage, which might have been better remaining as a mirage.

Unlike every other car, where the steering wheel returns to the centre, in the Mirage it wouldn't. So you'd just go round in circles. It was like trying to paddleboard without a fin. And you could see that it was a Japanese import because they added these terribly integrated switches for the UK market, like the rear fog lamp. Is there no fog in Japan? They literally just drilled a hole and stuck a button on. It was such a shambolic, shoddy car.

Mitsubishi made a grand tourer called the GTO (also called the 3000GT), which was full of tech, had four-wheel drive and a turbocharged engine. It was amazing on paper but it turned out to be rubbish compared to the superb Honda NSX, which also came out in 1990. The FTO was a cool-looking mid-engined coupé produced between 1994 and 2000 and the popularity of grey market imports led to Mitsubishi officially distributing it to several countries, including the UK, but it was right at the tail end of the production run, so a case of too little too late.

The Mitsubishi i-MiEV (first produced in 2009) was one of the first electric cars and it had almost no range, like about sixty miles. The Colt Car Company gave me one as a long-term demo vehicle and they offered to create a bespoke interior to promote the fact they were offering this feature to customers. I went with a matte black wrap with white leather interior, and they did the dash in white leather as well. It looked great, but I realised (after about ten seconds in it) that the angle of the wind-screen meant that the white dash reflected so much light that I literally couldn't see out. So I resorted to taking the black floor mats off and draping them over the dash when I was driving it. And people think that life as a motoring journalist is all glamour. Aside from that, which was kind of my fault, I really liked the i-MiEV. It's one of those cars that fits into that category I love of slightly weird, unusual cars, several of which I own, like my Citroën Ami Buggy, Fiat 126 and Suzuki Jimny. So even though in many ways the i-MiEV was crap, it was also kind of good, and I liked driving it.

In 2013, Mitsubishi produced the first decent plug-in hybrid, the Outlander PHEV, which had an electric-only range of twenty-eight miles (weirdly specific number but I guess the forty-five kilometre conversion is more of a round number) and really low emissions, which meant that you could have it for next to nothing as a company car. Sales of these things rocketed and they made a shedload of money out of them. But it was totally a one-trick pony. On that note, something unexpected I've learned about Mitsubishi is that they're really popular within the horse-enthusiast niche. Next time you see a horse box, check to see if there's a Mitsubishi pulling it – probably a Mitsubishi Shogun, a L200 pickup or an Outlander. And it wasn't just their customers, either. The Colt Car Company would hold a big annual event for the Badminton Horse Trials, which was only twenty miles away from where they were based, in Cirencester.

And that brings me on to one of Mitsubishi's most memorable offerings – the two-door rear-wheel- sporty coupé called the Starion (in 1982), but it wasn't the car that got people talking. It was the name. The story goes that it was supposed to be the Mitsubishi *Stallion*, but Japanese pronunciation problems meant that it was transcribed with an 'r' rather than two 'ls', which doesn't exist in Japanese. In Mitsubishi's sales brochure from May 1982, they state that the name comes from 'the name STARION, derived from the combination of star and Arion, Hercules' horse in Greek mythology, [which] symbolises a sense of the universe, and of power and high performance'.

I'll lay out the case for the prosecution and defence.

Stallion would make sense as a name, given Mitsubishi's previous horsey car names Colt and Eclipse (named after a champion racehorse) and the fact that it was pitched as a rival to the Mustang in North America. It also wouldn't have been the first naming disaster, seeing as Mitsubishi did come up with the car name Pajero (which came out in 1982). While the name came from the scientific name for the Pampas cat, *Leopardus pajeros*, one thing they hadn't factored in was that 'Pajero' in Spanish means 'w****r'. That's why it was renamed the Montero in North America and Spanish-speaking countries.

However, Mitsubishi had combined unlikely looking words together before for car names, like the 'Cordia' (a combination of the mineral 'cordierite' and 'diamond'), although lord knows why. And they did use Greek and Roman mythology as the inspiration for the names of engines, like the Saturn (produced from 1969) and the Orion (produced from 1977). Things are starting to look more positive for the defence. Well, they would have done if the Starion had actually featured an Orion engine. But it didn't. Hmmmm.

On YouTube, there's a Japanese TV ad for the Starion from 1982 that starts with the birth of a star and includes a logo of the Starion featuring the head, mane and shoulders of a horse. You could use that for either argument, so I'm none the wiser after two hours of research.

If it was a mispronunciation debacle, though, I do have some sympathy. A few years ago, I was staying in Chamonix and had a 3am start to make it to the Geneva Motor Show, where I was due to meet Mitsubishi Japan's top execs for an interview. When I arrived I was knackered and my brain hadn't cranked up to full speed yet. So of course I chose this moment to wheel out some basic Japanese. My attempts left the exec very confused and slightly amused. I turned to the translator.

'I'm saying how are you, right?' I said.

'No, no. You are saying excuse me, like you are apologising for, er . . . farting in a lift.'

And like a fart in a lift, that one lingered.

My top 10 small cars are: the 1957 Fiat 500 (*bottom*), Mini Mk1, Citroën 2CV (*above*), Ford Fiesta, BMW Mini Mk1, Renault Twingo Mk1, 2007 Fiat 500, Renault 5 E-Tech, VW Polo, Suzuki Jimny (*middle*).

My top 10 cars to drive on a country road are: Porsche 911 S/T (*top*), BMW E46 M3 CSL (*bottom*), Ferrari 458 Speciale, McLaren 675LT, Toyota GR Yaris Mk2, Mazda MX-5 RF, Lotus Elise Mk1, Alpine A110 (*next page, top*), Mitsubishi Evo VI Tommi Mäkinen, Caterham Seven 310S.

Alpine A110 (*above*) – basically a French version of a Lotus Elise but for the modern day. It has a punchy turbocharged engine and a terrific lightweight chassis and suspension that glides over the road much better than you imagined a sports car would.

Yes, the Boxer engine of the Subaru Impreza had a bit more character and noise, but the four-wheel-drive system in the Mitsubishi Evo (*above*) had this amazing yaw control system which meant that it handled better. As a result, I thought the Evo was the better car.

My top 10 cars to be driven in: Rolls-Royce Phantom (*below*), Toyota Century V12 (*above and middle*), Blower Bentley, Lexus LM, Rimac Nevera, Mercedes EQS, Skoda Superb 2.0 TDI, Range Rover Long Wheelbase, BMW 7 Series (G70), Audi S8 (D5).

The original Series 1 Land Rover (*above, right*) was introduced in 1948. It was a utilitarian vehicle and the basic design remained right up until the original Defender went off sale in 2016. Queen Elizabeth II loved them – Prince Philip even had a Land Rover Defender TD5 modified to carry his coffin.

The Mercedes G-Wagon AMG G 63 (*above*) was the perfect car for Covid because if we were facing the apocalypse, at least I had an amazing car to smash through all the zombies.

Above: Me with my friend, drag-race partner and car customiser extraordinaire, Yianni Charalambous. Alongside his Lamborghini Revuelto and an Audi RS 7.

Left: Drag-racing every generation of the VW Golf GTI.

Lamborghinis are named after fighting bulls – Aventador (*above*) was a feisty bovine that impressed the crowds at a 1993 event in Zaragoza, Aragon.

I like to do stupid poses next to cars, even priceless ones like the Mercedes-Benz C III-II concept (1970).

The 2.5-litre Mercedes 190E Cosworth (*above*) is a fast car. I remember one being able to keep up with my little Honda NC 30 sports bike when I was in my twenties. Not many cars could do that . . .

Driving and reviewing cars for a living is the best job in the world. (*Clockwise, from top left*): my Polski Fiat 126, Mazda MX-5 Mk2, Ferrari 458 Speciale, Bentley Continental GTC – painted for Pride – with my partner, Jo, and Pontiac GTO.

WORST CARS I'VE DRIVEN

1. **G-Wiz** – No wonder the Tesla Model S was an absolute revelation when most people's idea of an electric car was this. A lot of people had them in London to dodge the Congestion Charge, but it ran on an old-fashioned lead–acid battery and felt like you were driving around in an old milk float made of plastic. A horrible, horrible thing.

2. **Mitsubishi Mirage** – Absolutely dreadful. They had to convert it for the UK and it looked like said conversion had taken place in some bloke's shed. If you put it on full-lock and released the steering wheel, it wouldn't return to a normal position so you would end up driving in circles. For the rest of your life.

3. **Tata Nano** – This car was designed to get people in India off motorbikes and into cars. They brought one over to the UK, with the idea of maybe selling them over here. I drove one but it was too cheap, nasty and I just didn't feel safe in it.

4. **Perodua Kelisa** – Built in Malaysia, this was Britain's cheapest car in the early noughties. It did develop a bit of a cult following but it was cheap and nasty rather than cheap and cheerful.

5. **Rover CityRover** – Based on a Tata, but for what it was, it cost too much and felt like a half-arsed last-ditch attempt to sell anything they possibly could to try to make some money. A truly embarrassing moment for Rover. It would have been better to bow out gracefully.

6. **Ford EcoSport Mk1** – A confusing car in many ways. It was so badly built, looked ugly and I don't know why anyone bought one, which is why I was aghast when I found out my uncle had. It was like he hadn't listened to a word I'd been saying all these years!

7. **SsangYong Rodius** – Probably the ugliest car ever made – it looks like someone drove a van into the back of an MPV. Genuinely hard to take seriously. SsangYong have since rebranded as KGM and things are looking up because their cars are far less hideous.

8. **Suzuki Kizashi** – A car that no one really wanted. It wasn't terrible; it was just pointless. Suzuki's famous for making quirky, interesting cars but this was the epitome of boring. It's little wonder they sold so few.

9. **Chrysler Ypsilon** – A rebadged version of a Lancia. A crap car to look at and drive.

10. **Vauxhall Mokka Mk1** – This was dull even by small SUV standards and was a sign of things to come because it turned out to be a sales success.

Nissan

As a young motoring journalist in the noughties, I was handed the tasks that the senior journalists sensibly swerved. 'Mat, we'd like you to test the new Nissan!' my editor would say, with an encouraging upward inflection. There was no positive spin, however – Nissans were granny cars. What was so disappointing was that, not so long ago, Nissan had produced some of the most exciting cars around. It was like finding out that Mick Jagger had taken up lawn bowls.

As a teenager in the 1980s, Nissans were the hoverboards of the car world: they were seriously cool and way ahead of their time. The 200SX was that funky wedge shape with the pop-up headlights and the top spec had a 3.0-litre V6 engine. That was a wicked car, to use some eighties chat. You could pick up a 200SX dirt cheap fifteen years ago, but they've had a big resurgence in the drift scene, so they've shot up in price. And you can see why – with their highly tunable, four-cylinder turbo engines, real-wheel drive and limited slip differential, the 200SX was always a brilliant car for drifting. The 300ZX had less of a following, but it looked awesome with its slanted headlamps; and with its twin turbos, it was a properly fast car. And then there was the Bluebird ZX, which looked a bit like an Austin Montego, only much cooler. Plus, unlike the Montego, it didn't break, which is why every taxi driver from Tokyo to Torquay was driving one.

You knew the Skyline was going to be cool just from the

name. The first-gen Skyline was a successful touring car, but the early first- and second-gen production models weren't successful, partly because of the oil crisis of 1973, when the last thing anyone was thinking about were performance cars. Nissan revived the name in 1989 with the third-generation Skyline GT-R R32. A beast was born. With its 2.6-litre, twin-turbo, straight-six engine, all-wheel drive and four-wheel steering, it roared out of corners, winning pretty much every touring car race going on multiple continents. It was such a monstrous hit in Group A championships in Australia that they nicknamed it Godzilla. It was becoming a legend on the track, but it was also becoming immortal on the street. Thanks to its highly tunable engine, it was a modder's dream. Those four round tail lights were like the F1 starting lights. When they appeared, shit got real.

The R32 was also forbidden fruit, because you couldn't get one in the UK or the US. The Skyline GT-R R32 was never produced outside Japan and for a long time, the only export markets were Hong Kong, Singapore, Australia and New Zealand, but that changed in 1997. Nissan were permitted to sell the Skyline in Britain through one authorised dealer. But numbers were limited to 100 units. For modders, it was like Willy Wonka had just issued 100 golden tickets.

The next-gen model, the R33, was given a helping hand by the makers of the 1997 video game *Gran Turismo*, who slapped it on the cover, alongside a Toyota Supra. That game shifted nearly 11 million copies, becoming Sony PlayStation's bestselling game. Millions of teenagers around the world knew what a Skyline GT-R looked like. The next-generation Skyline GT-R, the R34, ticked every box. Engine, looks, performance: check. And this time, you didn't need to win a golden ticket to own one: Nissan officially imported it to the UK.

They ditched the Skyline name, after 2002, but there was no

doubt about the lineage of the GT-R (R35). When it roared out of the factory in 2007, a supercar slayer was born. At about 50 grand – half the price of a Porsche 911 Turbo – it was an absolute bargain. The engine was a 3.8-litre, twin-turbo V6 and it had a clever four-wheel-drive system, launch control and a racing car-style trans axle so the dual-clutch automatic gearbox was in the back, which gave the car perfect weight distribution. Computers controlled everything and yet the car still had this natural analogue-y feel to it, and that combination is a fine art. It felt like one of those times where Nissan had gone *all in* and in my experience, when a Japanese manufacturer opts for that approach, they create a spectacular machine. Like the Honda NSX. Like the Toyota GR Yaris.

The commitment to mastering a craft in Japan is on another level. When you have sushi in the UK, even in a nice Japanese restaurant, it's very good. You go to Japan and you imagine it tastes pretty much the same – I mean, it's raw fish, right. How much better can it be? Well, I found out on a night out with some of the guys from Nissan. I got talking to an *itamae* (master sushi chef) at a restaurant and found out that in order to acquire that title, they've trained for at least ten years. An *itamae* typically spends thousands of pounds on the tools of their trade – high-quality knives. He explained to me that the angle and precision of the cut preserves the integrity of the cells within the fish, and that skill is a blend of dexterity, anatomical knowledge and the quality and maintenance of your knives.

As much as I respected the man's obvious ability, it did sound like bullshit to me. He sensed that I doubted him, so he raised the stakes. He produced a beautiful, crisp headband, which would serve as a blindfold. Then, he beckoned one of his assistants over. They both prepared sashimi for me from the same piece of fish, and my job was to tell who had prepared which piece. The first one was incredible but the second even more so, tasting smoother and somehow sweeter. The *itamae* had made his point without any need for congratulation.

When the Japanese put their most highly skilled people on a job and give them the backing they need, what they create is pure, focused brilliance. When it comes to cars, there are some things that the European manufacturers are better at, like design flair and flamboyance. The Japanese are the masters of seeing what's already out there and perfecting it.

The GT-R was a welcome break from the series of uninteresting, uninspiring vehicles that Nissan (and Honda and Toyota) produced in the early 2000s, but it turned out that Nissan had something up their sleeve. They'd been working on a concept car since late 2002. It was conceived as a kind of baby brother to the X-Trail. That car, which became an absolute game-changer, was the Qashqai.

The Qashqai wasn't the first crossover SUV, but it was the first that was the right size, the right price point, well-built and fun to drive. Suddenly Nissan were surfing the crest of a wave. They couldn't make enough of them. I found myself answering the question 'What car should I buy' with 'Qashqai', irrespective of the person's age, budget and needs. That car impressed everyone. Nissan had gone from a struggling fringe brand in Europe to a forward-thinking innovator that could predict the future of motoring. Plus, they'd built them in Sunderland, which was a big win for the UK. In 2008, Nissan even started making them through the night to keep up with demand.

I think the UK and Japan are kindred spirits in some ways. We're both proud island nations and once we commit to doing something, we tend to do it. The Europeans have a different approach, and part of that is because they share borders so everything's a bit more flexible and everyone has to rub along. When you have to take a boat or a plane across a border rather than just amble across a vaguely defined line, it's a different deal. Having said that, there's a big difference between the Japanese and the British when it comes to mechanical reliability, and the Japanese have managed to instil that fabled reliability in the vehicles they've produced in the UK. Their plant in Sunderland is the most efficient car factory in the whole of Europe in terms of cars produced per worker. I remember doing an article for *Auto Express* walking around the Nissan factory in Sunderland. It took me forty minutes. In that time, twenty cars had rolled off the production line. They've made a car every two minutes for more than forty years. By the time you've finished this page, Nissan will have produced another car.

My first trip to Japan was to visit Nissan and I remember travelling on a coach that Nissan had chartered. The route took in some of the sights, including the Imperial Palace in Tokyo. I noticed a couple of old ladies in hi-vis jackets using long pincers to pick up litter. Seeing as you usually associate that sort of activity in that attire with enforced community service, I wondered what low-level misdemeanour these grannies were responsible for. Queue-jumping (actually a crime in a Japanese rail station)? Giving the wrong directions to a delivery person (also genuinely a crime in Japan)? Urinating in a park (surely not)? So I asked one of the Nissan representatives, who was quite taken aback. 'Oh no, they are not criminals. They have the honour of tidying up outside the Imperial Palace.' There is a lengthy waiting list for this privilege.

Another thing about Japan that you can't help but notice is that everyone has a job. If there's a pothole in the road, there will be a bloke wearing a hi-vis jacket and waving a stick. On the metro during rush hour, white-gloved *oshiya*, or 'passenger arrangement staff', push passengers on to trains and ensure that no one gets taken out by the closing doors. If you go to a car park, there will be a guy operating the barrier. They could automate a lot of their jobs, being the forward-thinking tech powerhouse they are, but they choose not to. It's partly because they want to make sure everyone's doing something to contribute to the community. But it's also because there are some jobs that rely on manual dexterity. I always imagined that when Japanese car manufacturers create a prototype engine, everything would be done by computer. A lot of it is done by hand, because our tolerances and sensitivity are finer.

When we arrived at Nissan's Tochigi Plant, they whisked us off straight away to go and race some GT-Rs. We learned that there's a dedicated Nissan plant where their finest craftspeople work. Nissan buy back used Skyline GT-R R34s, strip them and put them back together again, but better than before. What they're doing is creating top-end collectible versions of their very best cars. It's such a smart move. I got to drive one of these and, bearing in mind that a factory-default R34 is already an amazing car to drive, these customised R34s were astonishing. We're talking super-sharp handling, unbelievable responsiveness and motorsport levels of tolerance. When Nissan get it right, they create automobile alchemy.

Peugeot

In the past, Peugeot haven't resonated with me in the same way as Renault. Peugeot haven't produced many of those 'wow' cars; by and large, they've made a bunch of middling cars that do the job. And for that reason, I've never really wanted a Peugeot.

In fairness, the 205 was a great car. When you compared the handling and overall driving experience to a Fiesta, it was night and day. My dad almost bought a Peugeot 205 because we drove one on holiday in Paris and it impressed him that much. But in the end, as much as he wanted one, he couldn't justify the extra cost over a Fiesta or a Mini Metro, which is what we ended up with. Until I wrote it off.

The Peugeot 205 GTI was a stand-out car but I would have taken a Renault 5 GT Turbo over it all day long. Aside from the 205 GTI, not many Peugeots spring to mind. In 1991, they brought out the 106, which was a decent little car. The Rallye version of the 106, which was marketed as 'fewer frills, more thrills', looked great, as did the 106 GTI. That was an exciting car to drive and it filled the vacuum left by the 205 GTI, which had been discontinued in 1994. But from then on, everything went a bit samey and dull. Peugeot didn't stand out any more. They did a few OK cars here and there, like a GTI version of the 207 in the late noughties, but they were never at the top of the game.

They went off the boil with the successor to the 205, the 206 – it just didn't have the same presence. But they did pull it out of

the bag with the 306, which was brilliant. Sometimes a car manufacturer will come up with a car that looks right and drives right.
It's a difficult thing to do, but with the 306, Peugeot nailed it.
The 2.0-litre, 16-valve version, with the six-speed gearbox, was a
dream car in your late teens and early twenties. Sure, we dreamed
of Ferraris, but a top-spec Peugeot 306 would have done very
nicely.

Part of the problem is that Peugeot don't have a 2CV or
Renault 5 kind of car – a classic that everyone knows. There's no
sporty version of anything. No GTI any more. They also don't
have a car that everyone loved, which they discontinued and were
then able to spectacularly revive; like Mini did; like Fiat did; and
like Renault are doing. The issue with the 205 is that it just kept
going all the way from the early eighties to 1999 and then the
206 replaced it, followed by the 207 and then the 208. You can't
reinvent something that's already been reinvented several times.

Peugeot were one of the first, in the early noughties, to jump
on the folding roof bandwagon that Mercedes had popularised
in 1996 with the SLK. Peugeot produced their 206 CC (coupé
cabriolet) in 2001 and my partner's mum still has one, which she
bought new back in the day. The car's brain is completely fried
because its folding roof leaked, causing water to gather in the
footwell where the car's wiring loom runs. The upshot is that
the electrics go crazy and the doors lock when you're in it or out
of it. The car still runs OK, though! It's one of those cars that
you think you might be able to pour a bit more money into, and
finally fix it, but when you do that, a few months later, another
gremlin appears.

Peugeot did achieve something remarkable with the 3008,
the crossover SUV they launched in 2008. The first-generation
3008 was a truly hideous-looking thing, but the ugly duckling
turned into a swan with the second-generation model in 2016.

They might as well have called it something else, though, seeing as it bears absolutely no resemblance to the first-gen 3008.

The other thing they've excelled at is making their interiors far more interesting than any rival manufacturer I can think of. They designed a small steering wheel and lowered it so that, rather than looking through the steering wheel at the dials, the dials are above the steering wheel, so it feels more like a cockpit. Some people really disliked it because, depending on how you like to have your steering wheel and where you sit, it could be hard to actually see the dials. Many motoring journalists absolutely hated their new design direction, but if there's one thing I've learned, it's that when motoring journalists as a whole dislike something car-design-wise, it means you're on to a sales winner. BMW's recent big grilles on the M3 and M4 are good examples – all the journalists hate them, so they're definitely on to a winner. Peugeot took a risk but they did create something unique that people remember and I think the fact they were brave helped improve their sales. As did the overall quality of their cabins. They felt upmarket. If you were an alien coming from another planet and hadn't had any brand snobbery drilled in to you, based on the interior, you'd probably go with a Peugeot over a Mercedes. The trouble is, you don't buy a car just for the interior. I wouldn't buy a jacket just for the lining.

Motoring journalists focus on the sensations they get from driving a car, and often they don't care much what the car looks like. They're chasing that thrill. So while, with my motoring journalist's hat on, Peugeot's cars don't excite me dynamically, it changes when I look at them from the public's point of view. They've worked hard on their exterior and interior design, and it's paying off. Their cars look a lot more striking now. Their cars are a little bit different. They're quite practical and they've got decent engines. I add these factors up and they become cars that

I'd recommend to people. In 2021, we gave the E-208 the 'Best Small Electric Car' award and the 2008 'Best Small SUV' in the Carwow Car of the Year Awards. The Peugeot 5008 was highly commended in the Family Values category in 2025.

So, Peugeot make good, solid cars that I would recommend to the general public and normal friends and family in the market for sensible family transport. But would I recommend them to my petrol-head mates?

Nope.

Porsche

There's a meme online of me being interviewed in a podcast where someone asks: 'Lambo or Ferrari?' My answer's the same as it was back then: 'Porsche.'

If I was going to design a house, it would basically be a concrete box with big industrial windows and long sloping curves. Everything would be ordered, efficient and it would just work. Form and function coming together in harmony. Pragmatic but thrilling. That's Porsche. A Bauhaus building in car form.

A Lamborghini looks amazing, but they're a bit guilty of style over substance and that's something you only really discover when you're driving them. Also, I prefer something a little more understated, except for the 911 GT3 RS fun and games, which I'll get to. Also, unlike a Lamborghini, which feels like a car you'd drive occasionally, you can take a Porsche out daily. Unlike its rivals, Porsche plays this dual role of being quite easy to drive like a normal, everyday car, but at the same time, it'll thrill you. They make cars with a very wide bandwidth, but at the heart of it, Porsche make cars for people who love driving.

I was fifteen years old when our next-door neighbour's son, aged nineteen, bought himself a ten-year-old Porsche 911 SC. Porsches weren't that expensive in the early 1990s and he was living at home, had left school at sixteen and had saved up for it. It was that classic Porsche 'Guards Red' colour and big-tailed, and he'd reverse it out of the drive, stick it into first and the wheels would be spinning as he zoomed off up the road. Other neighbours were tutting and shaking their heads. Sometimes I joined in, in the spirit of good old-fashioned neighbourly grumbling, but I was really thinking, *That guy's cool, but that car is amazing.*

My godfather treated himself to the same model, the 911, in white. I remember squeezing into the back of it and seeing floor-mounted pedals for the first time. Unfortunately, he had some health issues and ended up needing an automatic, so he bought a Jaguar XJS. He missed his 911 so much. Each time he got into the XJS, he'd find something else that 'wasn't like my Porsche'.

I saw one Ferrari growing up in Walsall. They were mythical creatures but Porsches we'd see. Rich businessmen who liked driving cars would be in a Porsche 911. From 1975 to 1989, their top of the range 911 model was the 930 Turbo, which was so fast but notoriously difficult to handle, so it ended up with a cool nickname: 'the widowmaker'. If you could handle that car and remain alive, you were (a) a serious driver and (b) a nutter. That's an appealing combination for a teenage boy.

As a kid there was something ever so special about the famous whale-tail spoiler on the 930. It had these curved edges that looked so much better than the regular Carrera spoiler. And then there was the super-cool spoiler on the 959, Porsche's first hypercar. When it came out in 1986, it was the fastest street-legal production car, but even more than that, it was a technological masterpiece of a car with four-wheel drive and all sorts of crazy tech on board. Even by today's standards it was pretty blooming clever.

There's something about Porsche that makes me feel like they're designing cars – the 911 especially – specifically for motoring journalists, who generally love cars and like the sensation of driving. And to them, the Porsche 911 offers everything. It gives you a sports car, which is basically what every motoring journalist likes the most. We want it low, we want the suspension to be slightly firmer and we want a powerful engine that sounds good. We want the car to communicate with us when we're driving it, but we also want it to make sense. We also want it to be logical and user-friendly. We don't want to be asking, 'How does this work?' or thinking, 'This is annoying, this is uncomfortable, that's in an awkward place'. Porsche excel at attention to detail and the commitment to making something work, but they also add quirkiness, and I like that in a car. It fits my personality perfectly.

A 911 is a quirky car, because what other car has the engine in the back? Very few. A 911 has a unique feel to the way it drives. It has that beautiful flowing, soft shape on the outside but inside it's quite different. There's no extra trinket here or there, like BMW or French car manufacturers add – it's quite minimalistic and simple and beautiful, like a pebble on a beach that's been washed smooth of superfluous edges or imperfections.

When it was first revealed, in 1963, the 911 was being slagged off for having the engine in the back. But it worked and it even had rear seats. Sure, they're tiny, but you can fit kids in them (and car seats), so when it comes to the negotiation process of *Which car are we going to have* once you've had children, you can deny reality for a good while with a Porsche 911. 'It's fine – we can make it work. We don't need any more space than that!'

But there was a point when it looked like the 911 might not make it through the 1970s. Porsche introduced the 928 in 1977 to replace it. It was the first Porsche with a front-engined V8 and it was a different sort of car, designed as a grand tourer rather than an out-and-out sports car. But so many people still wanted a 911 that they've never got rid of it. A lot of people say that the VW Beetle is the most famous car in the world. For me, it's the 911. The Beetle's gone – people are forgetting about them. But the 911 is still alive and well. What Porsche mastered is automotive evolution. The 911 keeps getting better and better. There's a reason why a Porsche 911 of some sort has won twelve of the twenty-seven *Evo* magazine Car of the Year awards.

I was speaking to a friend the other day, and he was asking me about getting a big, expensive electric car for the salary sacrifice scheme at work (the employee sacrifices part of their gross, i.e., pre-tax, salary to cover the lease payments for an EV), so we were talking through the numbers. People don't just come to me for motoring advice. I am a trained chartered accountant, after all. Then my friend sighs and says:

'One day, I'll have a 911.'

'Have you got the money to buy a 911 now?' I ask him.

[Pause] 'Yeah.'

'You know what you're going to lose on the salary sacrifice deal in terms of take-home pay, right? You'll have the car on a two-year lease so multiply the salary sacrifice by two. If you compare that figure to you buying a brand-new 911 now and selling it in two years' time, you will lose less on the 911 than you will with the salary sacrifice scheme. You want the 911. It's cheaper to buy the 911. So buy the 911.'

There's always some maths that I'll bring out to justify owning a 911. And that's why I've got two, the cheapest one and one of the most expensive.

The 911 996 (the fifth-generation 911 produced between 1997 and 2004) is the cheapest but I bloody love that car. As for my other 911, well, I'm actually writing this on the day that I picked it up and unwrapped it. My god, what a car that is. The 911 S/T is the lightest and most beautiful Porsche 992. It's got a redesigned grille at the back for the engine, magnesium alloy wheels, new carbon-fibre wings and a new front bumper, but it's all about that 4.0-litre, naturally aspirated flat-six GT3 RS engine. It's the best engine in the world. The car around it is brilliant but I've bought this car for that engine.

In the 911 S/T, you can opt for the heritage design pack, which costs around £16,000 all-in and gives you a premium two-tone leather interior, jazzed-up dials, a gold 911 plaque on the dash, 911 lettering on a light silver door sill guard, Porsche crests on the seats, special coloured wheels and various other luxury bits and bobs. It's something dreamt up by the marketing department at Porsche and not something the chief engineer in charge of the GT division was very impressed by, mainly because it all adds a little bit of weight. But sometimes the marketing guys are bang on. The very best sports cars are thrilling at all times: when they're stationary, when you're driving them slowly and when you're driving them fast. The heritage design pack gives you all that.

I'm not so much of a fan boy that I won't criticise Porsche for things that annoy me but I honestly don't think they mind because they're so confident about their products, and they have a right to be. You can tell when a manufacturer knows how good the car they're launching is because they're genuinely excited about it. And they'll take any criticisms that you have and respond with good answers or explanations. They might even change the feature you're grumbling about.

Sometimes the Porsche marketing guys do come up with some rubbish, though, like the puddle lighting feature on the S/T. So, when you open the door, a little light shines from underneath the door mirror and displays a logo on the floor. Normally on a Porsche, the puddle lighting has the Porsche logo. Lovely. But because the S/T is celebrating the sixtieth anniversary of the 911, they've come up with this slogan in a flowy script that reads: 'Icons of Cool'. This is a classic example of trying to be cool but not understanding that mentioning 'cool' negates the coolness.

So, of course, I took the piss out of them for being cheesy and spectacularly uncool. Coincidentally, the UK press office removed the 'Icons of Cool' puddle lighting feature from the press 911 S/T and replaced it with the regular, lovely written script Porsche logo from a normal 911.

To their credit, the Germans know they're not always the coolest and, just like the Brits, they're able to laugh at themselves when they've had a shocker. And they have had a few of them. For example, Porsche haven't always been the savviest business operators. While other manufacturers ditched air-cooled engines long before, Porsche stuck with them until 1998, and I loved that about them. The trouble was they were making these engines by hand, which was costing a fortune, so they had to bring in Toyota to tell them how to set up a proper manufacturing process.

And then in the early noughties, Porsche found themselves in a position where they were looking to buy VW. The companies are all interlinked anyway – Ferdinand Piëch, chairman of VW's executive board, was Ferdinand Porsche's grandson and VW manufactured many of Porsche's vehicle parts. Porsche started to publicly buy up shares in VW (using debt) but they were also *secretly* buying options on Porsche shares, as well as going to great lengths to conceal their debt. When it came to light in October 2008 that Porsche effectively controlled 74 per cent of VW's shares, VW's share price went into overdrive. Suddenly, it was the most valuable company in the world. Ultimately, Porsche's spiralling debt during the financial crisis of 2007–2009, when banks were calling in repayments, meant that it couldn't complete a takeover. But it got worse for Porsche. They seriously needed cash, and they got it when VW bought a 49.9 per cent stake in Porsche in October 2009. Over the next three years, they kept nibbling away, and in July 2012, VW acquired the remaining

50.1 per cent. It was officially a merger, in the same way that me eating a burger can be seen as a merger.

Porsche are having a tough time at the moment, though. At the time of writing, sales aren't as strong as they have been. As reported in the press in 2024, Porsche allowed the contracts of 1,500 fixed-term employees to expire and in 2025, they announced that a further 1,900 jobs in Germany would go by 2029.

As for their cars, the Taycan's taken heat from the media, but unfairly so. During the first big wave of electric cars, the Taycan became the most desirable electric car on sale. Everyone was buying electric cars through their employers or businesses, partly because they're tax deductible. They were also early adopters who like to embrace the new tech. So Porsche sold shedloads of Taycans during Covid but then people's finance deals or leases came to an end at the same time so there was an avalanche of Taycans on the used car market. Early adopters were willing to pay more than £100k for a new Taycan, but on the used market, in which values reflect what people are willing to pay with their own money, Taycans don't hold their value like the more desirable petrol-powered 911, Cayman or Boxster. Added to that was the hesitation about buying electric cars because of the charging faff and range issues. Plus, people are wanting to have one last hurrah with an internal combustion engine car before the ban on ICE cars comes into effect. All that means that Taycans have depreciated but no more so than the internal combustion-engined Panamera. But in the current climate of electric car bashing, the Taycan became the target. I see it as a victim of its own success. There are so many of them around because it was so successful. It's a massively impressive car.

When it came to making a hybrid version of their hallowed 911 (the GTS T-Hybrid), Porsche fitted a tiny battery on it, and the battery is just there to drive an electric motor in the gearbox to add performance, to fill in the torque gap. It also powers a little electric turbocharger, which can spin up on electric power and then run the normal way after it's spun up, so the car only suffers from minimal lag. The effect of this is that, while other hybrids feel like hybrids, the Porsche doesn't – it feels natural. Porsche have managed to make a turbocharged car feel more like a naturally aspirated car, but with the torque of a turbo car. That sums up Porsche. Yes, the 911 GTS T-Hybrid is quite expensive but it's so impressive, responsive on the road and fast – noticeably faster than the old GTS (the 992.1). Porsche wanted to make it worthwhile to have a hybrid and they've nailed it.

It's going to be mad when they release the 911 Turbo S hybrid, which will be a crazy car. I've been invited to the unveiling in September 2025 and I imagine the car will go on sale towards the end of the year. The 911 Turbo S was the Carwow drag-race king for so long, just because it launches so well. It was beating Ferraris, McLarens, everything. All day, every day, it was delivering the numbers. It finally got beaten by later McLarens and later Ferraris. But when the 911 Turbo S hybrid comes along, it's going to be a sub-10 second car for a quarter-mile, I know it. The Porsche 911 GTS does a 10.8-second quarter-mile. I think the new Turbo will do a 10.4 and I think they'll get the Turbo S under ten seconds.

I do like every single Porsche. I've already mentioned my 911s, but I've also got a Boxster 986, which I bought blind at auction for five and a half grand a few years ago and is such fun to drive, and a 2007 Cayenne Turbo that I bought for seven and a half grand. That Cayenne is so fast that I beat Yianni in a drag race against a brand-new 4.4-litre, twin-turbo V8 Range Rover Sport. He had more horsepower than my Porsche. But I had Porsche horsepower. It's the same story – the Porsche has less power but it still wins. Their torque curve is better, their track control is better. To be honest, I don't know exactly what it is. One day I'll find out.

And then there's my recently departed GT3 RS. There's nothing else like that car – it's an absolute weapon and I loved it at first sight. But there's no excess to it, which sounds strange when you first see it because it looks like how I would have drawn a sports car as a child. But everything has been designed to make the car as fast and as stable as possible. They put a massive wing on the back, to generate the maximum downforce, so the engineers needed to do a lot of work on the front of the car, like moving the radiators to the front boot. Unlike Mercedes and BMW, who often feature fake, cosmetic vents on their cars, you won't find a fake vent on a Porsche. Everything that's on the GT3 RS is there for a reason.

I planned to keep my GT3 RS and did a lot of content on it but in the end I sold it because I wasn't using it as much as I thought I would. I do love the racing-car-like magnetic paddle shift, with the magnet pulling the paddle and making a clicking sensation. It doesn't make the gear change any faster but it adds a layer of interaction with the car that's both cool and somehow acts as a stress reliever. I tried them in the GT3 RS press car and ended up buying the exact same spec as the press car. I didn't go to the launch of the GT3 RS at Silverstone – instead, I asked Porsche to send me the car so I could spend more time with it on our test track. Which they did.

Believe it or not, I didn't used to have a great relationship with Porsche, and I think it was because at first they perhaps saw Carwow as small fry. But as we put out more and more good content, they started to respond better. Plus, we started to get a tonne of views, which always helps.

Fast-forward to the launch of the 911 S/T in 2023 and I see Porsche's head of PR for the 911 coming over to me: 'Mat, we want to do something a bit different for the launch of the new GTS [Porsche's hybrid 911]. We'd like you to do a drag race, pitching the old car against the new car. We want you to race Mark Webber.'

Stepping out of the car, I couldn't help but think back to me as a boy lusting after my next-door neighbour's 911. And then, here I was, alongside Mark Webber, as part of the launch for a landmark 911.

BEST HYPERCARS OF ALL TIME

1. **Porsche Carrera GT** – This is probably the best-sounding car ever made, plus its incredible V10 engine is mated to a six-speed manual gearbox. Just be careful because the clutch is easy to burn out and expensive to replace.

2. **Koenigsegg CC850** – It has a crazy 1,185hp twin-turbo V8 engine but the best thing about it is its mad nine-speed automatic gearbox, which can double as a six-speed manual. I haven't made that up. Bonkers yet brilliant engineering.

3. **McLaren F1** – Created by the McLaren Formula 1 team's designer Gordon Murray and powered by a BMW V12 engine, it does a top speed of 240mph (verified) and held the record for the fastest road vehicle for eleven years.

4. **Gordon Murray T.50** – Designed by Gordon Murray, this has a naturally aspirated V12 engine, a manual gearbox and weighs only 997kg. It's an insane car.

5. **Bugatti Chiron** – Following on from the brilliant Veyron, but it has 1,500hp in the standard form and 1,600hp in the Super Sport.

6. **Pagani Zonda R** – A track-only version of the incredible Pagani Zonda powered by a naturally aspirated V12 engine from a Mercedes racing car. It is automative art combined with brilliant engineering.

7. **McMurtry Spéirling** – This car can defy gravity with its crazy fan that sucks it to the ground. It gives you

Formula 1 levels of downforce at zero miles per hour, which means you can corner faster than any other car. Even novice drivers can corner as fast as Max Verstappen in this car.

8. **Ferrari F40** – Ferrari's fortieth birthday present to itself and what a gift. Crazy twin-turbo V8, insane looks and absolutely delightful handling. I want one.

9. **Mercedes-AMG ONE** – Powered by the same engine that won Nico Rosberg his Formula 1 World Championship. It really shouldn't work (and often doesn't) but when it does – what an incredible car. A true collector's piece.

10. **Aston Martin Valkyrie** – Like a Formula 1 car for the road because it's been designed by ex-Red Bull design guru Adrian Newey. Powered by a naturally aspirated V12 that screams like a banshee.

Renault

My parents had French friends when I was growing up. It started when a French family came to stay with my grandmother as part of an exchange trip and they got on very well with my dad, who was a bit of a Francophile. So we became friends with this French family. We didn't tend to go on holidays to hotels and that sort of thing. We'd always go and stay with folk. These folks. They'd always be driving a Renault and their *voiture* seemed slightly cooler, quirkier or just better than the equivalent British car. Both sets of kids were jammed into the back of this Renault 25, bouncing around with no seat belts, with the windows open, in France. We were having the time of our lives.

In my early teens, it was all about hot hatches. And it was the French car manufacturers who were producing the best ones. A lot of my friends thought the coolest one was a Peugeot 205 GTI, released in 1984, but there was one that topped it for me: the Renault 5 GT Turbo. The shape of it, the turbocharger, the sound it made, the fact that it went like stink, everything. It could do 0–60 in about 7.5 seconds, which was mental back then. When I saw one, I just used to stare it at longingly. I still want a Renault 5 GT Turbo now, but they've become so expensive.

The next seriously impressive Renault hot hatch was the Clio Williams that came out in 1993, when I was up at Edinburgh University. The Clio Williams was designed for rally racing but became a huge hit on the roads. It was named after the Formula 1 collaboration between Renault and the Williams team, and that was pretty awesome in itself given that Williams-Renault were the reigning F1 constructors' champions and Williams' Nigel Mansell was the world champion driver in 1992. Despite what they say on the Renault website, the Williams F1 team had nothing to do with the Clio Williams – that was down to Renault Sport, the company's motorsport wing. The Clio Williams did have an F1 connection, though – it became the safety car for the 1996 season.

The Clio Williams looked the part, with its gold rims, gold Williams lettering, bulging beast of a bonnet and 'metallic sports blue' paint. The first 3,800 cars they made each featured a gold plaque on the dash with the Williams and Renault logos, which was even cooler. Of course, given the F1 'association' and the fact it was an incredible car to drive, they sold out immediately, so ended up producing another 1,700-odd. The car had a 2.0-litre, four-cylinder engine that generated 145bhp, taking you from 0–60 in around 7.6 seconds. They made another two series of the Clio Williams, producing around 12,000 of them, although many of these were destined for the race track.

The Clio Williams was an incredible hatchback, but nothing topped the Renault 5 GT Turbo. The Renault 5 just looks *right*, still, and that's why it's been so successful and has taken Renault so long to revive it. Strangely, the shape of it reminded me of a butter dish with a lid that our French friends would lift off theatrically. To be fair, when you make butter as well as they do in France, it needs a bit of theatre.

The new Renault 5, released in early 2025, is Renault's equivalent of the new Mini in 2001 or the reinvention of the Fiat 500 in 2007. Only, this one's electric. It's a renaissance for Renault because they've become cool again, after years of focusing on worthy, decent cars that were good but just not particularly desirable. The new Renault 5 has given them their va-va-voom back.

It's great to look at, great to drive and reasonably priced. Yes, it's a bit cramped in the back, but that's my only criticism for an otherwise brilliant car that I'd happily own.

I genuinely think that Renaults are decent, good-value-for-money cars, and the majority of them I've reviewed well. Then there was the Renault Scenic E-Tech. I also had the Peugeot 3008 at the same time, which is one of the Scenic's competitors. For roughly the same money, the Peugeot just felt like the more premium car, and when this happens in a direct head-to-head you do tend to think, *Why would I have the other one?* So, when it came to grading them, I gave the Peugeot four out of five, which meant I kind of had to give the Scenic three stars. Most of the other motoring journalists gave it a safe four stars. And it even won European Car of the Year, which Renault pointed out to me after they'd seen my review of the car in the video.

It's not uncommon for manufacturers to question your verdict on a car, especially when the car in question has picked up a gong. This tapped into a long-running issue for me, though. I can't count the number of times when the European Car of the Year award has been announced and I've reacted with: 'How the hell has that won?!' In 2001, the Alfa 147 beat the Ford Mondeo, and that decision was crazy. In 2002, the Peugeot 307, which wasn't all that, beat the new Mini, and that car was incredible. The Chevrolet Volt beat the Range Rover Evoque in 2012, which was ridiculous, and the Volt went on to sell terribly. Peugeot have won three times since 2014, and each of them were good cars, but they weren't winners – not in my book, anyway.

Sometimes a manufacturer is just not going to agree with a verdict you've given, which is what happened with the Renault Scenic E-Tech. Other times you can give a fair verdict on a car but you structure the video all wrong so the positive message gets obscured. I think that's what happened with my review of the

Dacia Spring (Dacia are owned by Renault). What I was trying to say was that it was a fit-for-purpose budget electric car, but my positive message about it was right at the end of the video, so may not have landed with the desired effect. The timing of the video was also unfortunate because we'd just decided to experiment with some new content formats – you have to evolve on YouTube or you soon get forgotten – to keep things fresh. The Dacia Spring happened to be the first video in that style. The title of the video and the fact that I'd started with all the negative points set the wrong tone and meant that my verdict was perhaps misunderstood by Renault.

In a meeting with the then Renault UK MD, I held my hands up and acknowledged that I'd got it wrong. He took it in good humour and was quite a cool guy about it. Unfortunately, two weeks later, my video of the Renault 5 E-Tech came out. That was the one where I couldn't help but focus on the car's woven wicker basket to the left of the glove compartment. This sounds like an April Fool's joke, but it's genuinely a baguette holder. I suggested it might be more useful for holding something rude. The Renault MD did seem to see the funny side again, though, and sent me a baguette in the post with the note: 'This is what it's really for.'

Then things took a turn for the worse. In March 2025, I shot a video pitching a Dacia Duster in an off-road battle against a Toyota Land Cruiser. It was a deliberate mismatch. The Duster is a £23k SUV and the Land Cruiser's a £74k beast famed around the world for being a brilliant off-roader. But the Duster's a capable car and in all but a couple of the challenges we set, the Duster (which I was driving) kept up with the Land Cruiser, and the final scores were nine each, although I awarded the moral victory to the Duster. However, in one of the earlier challenges, when I was driving along a very bumpy track, I hit a huge bump on the front right and the suspension (and my back) took an absolute walloping. There was a nasty smell and it wasn't just coming from my trousers. It looked like we'd have to buy the car off Dacia (although we checked and the car wasn't actually damaged). Because we checked the car and it looked fine, we forgot to tell Dacia about the incident. So the first they saw of it was in the video. To make matters worse, after the bump in the video, I joked that the marketing exposure for the car would be worth more than the cost of any repairs. Needless to say, Renault didn't see the funny side. As for the car, it's a very decent machine for the money – so much so that we awarded it the Carwow 2025 Car of the Year in the Smart Spender category, partly based on its performance in the video.

It's been funny putting this Renault section together because it's made me realise how many Renaults I've really liked and how big a fan I am. For example, one of my favourite hot hatches ever is the Renault Sport Mégane R26.R. I love love love that car. The second generation was brilliant as well, but then it started to get a bit too expensive and the Honda Civic Type R overtook it, but I'm still really fond of it. I'd like to buy a Renault Avantime, their coupé MPV that was based on the Espace and developed by the same guys – Matra – who conceived and built that. Matra were a company that mainly built racing cars and competed as a constructor in Formula 1 from 1967–1972. They had some pedigree too – Jackie Stewart won the 1969 World Drivers' Championship in a Matra-Ford. Matra's partnership with Renault was a commercial success and the Avantime was a great-looking car – unusual and quirky, but cool. During that period in the early noughties, Renault experimented with their designs and many of them really paid off.

One of them was the Renault Clio V6. Utter madness of a car. As was the Renault 5 Turbo 2, which looks amazing. Renault are currently making a limited-edition EV version of the Renault 5

Turbo 2 with 500hp and a price tag of around £150k. Are people going to want it, though? They're taking a big risk because electric hypercars and special-edition models are plummeting in value and are hard to sell. So, as good as it looks, I'm not sure about it. If they'd put in an internal combustion engine, it would be a huge hit, but they're focusing on EV production to hit the EU's zero emission vehicle mandate.

The Renault Sport Spider was a late 1990s roadster that was a bit of a flop, but it just looked amazing and I really wanted one. It was another mad offering – it had no roof, it didn't drive that well and it was completely impractical, but it was so cool. The Renault Twizy, their two-seat electric quadricycle that came out in 2012, was another cool offering but the tandem seating wasn't great. It was, however, ahead of its time, and that's something Renault continued to be in the 2010s. They were quick to the EV game with the Renault Zoe, which is achingly boring-looking but really a good-value, decent little electric car, although they did that weird thing where you had the option of buying the car but leasing the battery, because the battery was quite expensive. There are still cars out there now that you could buy on the used market where you don't actually own the battery; you pay a lease on it. But as for Renault, they were forging a reputation for being quite pioneering and innovative, trying new things and taking the risk that some-times you get burnt. I like that attitude.

When I started as a motoring journalist, Renault were one of the manufacturers who were really pushing improvements in car safety. Everyone thinks about Volvo, and rightly so, but in 2001, the Renault Laguna was the first car to be awarded five stars by Euro NCAP (the European New Car Assessment Programme). The five-star rating had only been created by Euro NCAP less than a year before to reward a car's ability to perform well in front and side impacts. Some manufacturers thought there was no chance they'd ever meet the new stringent standards, but within a year, the Renault Laguna smashed it (you know what I mean).

I'm going to end the Renault chapter with a story that Renault will probably read as *Mat Watson abuses Renault to start criminal enterprise*, but here goes. So I had a Renault 5 for a while in the noughties. When I was at *Auto Express*, I met a friend of a friend who'd been living in London for years and the whole time he'd been driving around a French left-hand-drive Renault 5. I asked him why he didn't get a right-hand-drive one and he told me: 'Because I get away with anything in it. Parking tickets, congestion charge, whatever. They don't bother sending tickets over.'

I didn't believe him. So I put it to the test. I went over to France, bought a cheap Renault 5 and properly registered it with my French friends at their home address. Once I'd done all of that, I drove it back to the UK. I didn't want to do anything illegal, so I didn't speed, but I did go out of my way to accumulate a full house of civil infringements: parking tickets, congestion charge tickets, yellow box encroachment tickets, bus lane tickets – all that shit. Six months went by. Not one letter came through the post. I even took the car on a little trip around the UK – to London, Birmingham and Leeds, and scoped out where the traffic wardens were before deliberately parking on their route. I jumped out and loitered around nearby to see what the traffic wardens did when they approached the car. All of them did the same thing. They paused, looked puzzled for a minute and then walked past it. This story got picked up by the national press because it revealed how people on foreign plates were avoiding fines.

I sold that Renault 5 to the French guy who gave me the idea. Whenever I see a parking warden looking confused, I wonder if I might catch sight of my old Renault 5 again.

Rolls-Royce

If you wanted to advertise that you were a millionaire in the 1980s, you drove a Rolls-Royce Silver Shadow. Or better still, you had someone else drive it for you. We can all picture the owner: silver hair, smart shirt, sun tan, posh voice, gold watch, red trousers.

That image is quite a contrast to the state of the company ten years before: arse hanging out of their trousers after all their cash was sucked into their RB211 turbo-fan jet engine. The British government rescued the company in 1971, and the engine that crashed Rolls-Royce also propelled the revitalised company into the world's leading aero-engine producer. The automotive side of Rolls-Royce was separated off and sold to Vickers in 1980. In July 1998, VW bought the business for £430 million, giving them Bentley (part of Rolls-Royce since 1931), the factory in Crewe, the iconic Rolls radiator grille shape and the rights to the legendary Spirit of Ecstasy bonnet mascot. But they'd soon have some agony to go with their ecstasy. The deal didn't include the rights to the Rolls-Royce brand name or the logo – those trademarks still belonged to the aerospace wing of the company. VW could now make a Rolls-Royce, but they couldn't call it a Rolls-Royce.

Just a month later, in a deal done on a golf course that sounds like a scene out of *Goldfinger*, BMW bought the rights to the Rolls-Royce brand and logo for £40 million. It looked like BMW had pulled a rabbit out of a hat. VW were left holding a very expensive hat, which for legal reasons they weren't allowed to call a hat. In what I think I'd struggle to see as a gesture of goodwill, BMW allowed VW to keep using the Rolls-Royce name until 31 December 2002. After that, the new company, Rolls-Royce Motor Cars Ltd, would take over. And they didn't hang around. They unveiled their first new Rolls-Royce – the Phantom – at 00:01 on 1 January 2003. They weren't done with the big reveal; they also officially opened their 42-acre Goodwood plant in West Sussex that day.

So much was at stake when the new Phantom was launched. Bentley were outselling Rolls-Royce by ten to one at that time and Rolls-Royce were gambling everything on the most expensive production car in the world. If it failed, this would be the last Rolls-Royce. But what a car the Phantom turned out to be. It had such road presence, such opulence and no old-man whiff about it whatsoever. You couldn't take your eyes off it. It looked like something out of Gotham City.

The Phantom was an astonishing feat of design. It was big, bold, in-your-face, and with no expense spared to achieve the utmost luxury. Remarkably, it still felt like a very British product, even though many elements were German. Final assembly was done by hand at Goodwood and you really got a sense of the intricacy of the hand-crafting once you ran your hands along the beautifully made upholstery and thick wood veneers. The starry night feature (which they call the Starlight Headliner) on the inside of the roof was a dazzling touch. It's made from 800 fibre-optic lights that shine through minute perforations in the roof lining. Each of those stars has been placed by hand and it's not just a random scattering of stars – it depicts the constellations above the Goodwood factory on the night of the Phantom's launch. But you don't have to go with the default. You can ask them for the night sky on the date of your kid's birthday, the date your football team won the FA Cup, whatever. It's fair to say that Rolls-Royce are decent on details.

The iconic pinstripe running along the side of the Phantom – and all Rolls-Royces made since 2003 – are hand-painted by one man, Mark Court, who uses squirrel-fur brushes that he trims himself (the brushes not the squirrel). He even powders his fingers so they glide along the surface as he's working. He gets flown around the world with his little toolkit to paint the stripe on for customers. It's high-stakes work because it's not like you can rub the paint out with an eraser. Once the paint is on, it's on.

The Phantom's powerful 6.75-litre, twin-turbo V12 engine was originally a monstrous 9.0-litre V16. They made a few of them, one of which ended up in the car used in the second *Johnny English* film as a special request from Rowan Atkinson. My most-viewed video on TikTok is me with a pound coin, which I stand up and place directly on the Phantom engine. Even when I revved the engine, the pound coin just rotated slightly; it didn't fall. That's unbelievable.

The Maybach 57 came out around the same time as the Phantom and they're both excellent cars, but the Rolls feels way more expensive inside. It's like you're driving around in a drawing room. Felix Dennis, the guy who owned Dennis Publishing and *Auto Express* owned *both* – the Phantom and the Maybach 57. He preferred the Maybach over the Rolls for one reason: the ashtrays were bigger. Anyone else in the world would take the Phantom. The Maybach 57 is a first-class plane ticket but the Phantom is a private jet.

There's something about travelling in a Rolls-Royce with its air suspension that makes you feel like you're serenely floating along. You enter a sort of meditative state when you drive it, caressing the car along the road with your fingertips on the delicately thin-rimmed wheel. Problems, fears, nuisances and unnecessary noise belong in the outside world, not in a Rolls. I'm not sure if this is marketing bullshit or not, but when Rolls-Royce created the second-generation Ghost, it was apparently so eerily quiet inside that the motion that people's inner ears detected wasn't matched by their other senses. And this had the effect of freaking people out and making them feel nauseous. So Rolls-Royce re-engineered a little bit of noise. I would have loved to have heard that conversation at Goodwood. 'I'm afraid you've messed up, fellas. The car's too good.'

The iconic grille design of the Rolls-Royce does look like an Ancient Roman temple, and that's no accident – it's inspired by the Pantheon in Rome, which has wider columns at the edges and narrower ones towards the centre to give you the illusion of symmetry when you look at it from the front. And whereas the grille on a Bentley Continental is plastic disguised to look like metal, on a Rolls-Royce it's metal. The Spirit of Ecstasy bonnet mascot now pops up when the car is turned on. And if you try to grab it, a little trap door opens and it disappears down into the bonnet. The latest Spirit of Ecstasy, on the Spectre, has a slightly lower stance, making her more aerodynamic and probably giving you another 100 yards of range. Rolls-Royce spent 830 hours designing it and testing it in a wind tunnel.

Unlike every other car, which has a rev counter, a Rolls-Royce has a power reserve meter, measured as a percentage. Whereas a rev counter tells you how hard you're working the engine, a power reserve meter lets you know that you're never going to stress this engine out, so there's no reason to get worked up. Like the bank accounts of Rolls-Royce owners, there's plenty left in reserve if one should require it.

Choosing extras on a Rolls-Royce is ever so slightly differ-ent from the options on, say, a Ford Focus. With other car man-ufacturers, the press office will have a document listing their various options and the prices. But with Rolls, like a menu in a fancy restaurant, there's no mention of money anywhere. Perish the thought that a £ symbol would influence someone's decision-making. Also, Rolls-Royce don't really have a conventional options list. They'll do anything you want. You can select the exterior paint from their 44,000-colour palette. Or you can ask them to come up with a bespoke colour, just for you. If you'd like them to fell, shape, smooth and polish a tree on your estate for your interior wood panelling, probably just ask.

When Rolls-Royce launched their first electric vehicle, the Spectre, my first thought was that I was getting to drive a real-life version of Cruella de Vil's car from *101 Dalmatians*. It was the same plum colour and the hood was the size of a driveway on a country estate. At first, it did feel like an electric model suited Rolls-Royce. They're both silent, effortless, pure, untarnished. But Spectres aren't holding their value and the people who have bought them aren't emphatically keen about them. It's nothing that Rolls-Royce have done wrong. It's more about the nature of the EV landscape. All electric motors are now very quiet and super smooth. The Chinese can build a powertrain as smooth and refined as Rolls-Royce can. What they can't do is construct a V12 engine that is as smooth and quiet as an electric motor. The unparalleled engineering, design and craftsmanship needed to make an internal combustion engine that quiet is part of Rolls-Royce's USP. In that sense, a Rolls-Royce is like a Rolex. Anyone can make a digital watch that keeps time just as well as a Rolex, but that's not a measure of exclusivity. The intricate skill, knowledge, patience and commitment to excellence required to achieve mechanical perfection is a showcase of human endeavour. And it all comes to life when you hear that 'tick'. It's the same feeling when you place a pound coin on a running Phantom engine and it stays there.

I find it harder to get cars from Rolls-Royce than I do from BMW. Although it's all the same company, Rolls-Royce have a different press office, and they're a different sort. They're also very specific about what you can do with their cars. So when I do manage to get a car out of Rolls-Royce, I won't be allowed to put it head-to-head with a rival car. The subtext is *we don't compare ourselves to others. We're Rolls-Royce.*

SEAT

I sense that SEAT are a problem child for VW and they don't know what to do with them. When you think about VW's brands – Audi, VW, Skoda and SEAT – each of the others has a distinct identity and appeal.

Sporty and plush: Audi.

Solid, middle-of-the-road option: VW.

Budget option that's full of surprises: Skoda.

Trying to explain SEAT's position in the market among the VW brands is more long-winded. They're good value at similar money to a Skoda. They've got the same engines as VW so they're solid and the styling is sportier than equivalent VWs, but they're less expensive and not quite as desirable. You can see why they've got a branding issue.

But if you can get past that, SEAT make good cars that I like and rate. Their shapes and designs are often a bit more interesting than VWs. Even though buying a Skoda probably makes more sense to more people, because you feel like you get more car real estate for your cash, I prefer SEAT because they appeal more to people who like driving. And for these reasons, when my mum asked me to point her towards a good all-rounder that's decent value for money, I chose the SEAT Leon. And she was very happy with it (until I got her to sell it for a Tesla Model 3). There have been SEATs that I'd recommend to friends, like the fourth-generation SEAT Ibiza (produced from 2008) with the 1.2-litre engine, because I knew the engine was the same as the one in the VW Polo Mk4, which was quite punchy, albeit a bit unreliable. I would have taken the Ibiza over the Polo back then – they were both very good, similar cars – but you'd be paying extra just for the VW badge.

One of the problems is that SEAT haven't nailed their marketing in the way that Skoda have done. Skoda had a story. They were crap, then got taken over by VW, owned their crapness from the past and turned it into a brand that people admire. SEAT don't have a story. Well, they do, but it's a crap one. The company started as a joint venture between the Spanish government, Spanish private banks and Fiat, who they basically relied on for technical assistance. Yawn. It's only when Fiat withdrew from SEAT in 1982 that things started to get more interesting.

SEAT's first offering as an independent company was the Ibiza. And they pulled together the big ones for that one. Giorgetto Giugiaro, the legendary car designer behind the DMC DeLorean and the Mk1 VW Golf, was brought aboard for the exterior design. Renowned interior designers Karmann, who'd done the VW Golf Cabriolets, the Scirocco and the VW Karmann Ghia, did the insides of the Ibiza. Porsche helped to develop the powertrain and they allowed SEAT to inscribe the camshaft cover of the Ibiza with 'System Porsche' for a royalty of seven German marks per car. This was a good time to collaborate with Porsche seeing as they were down on their luck and looking for business opportunities. There was a SEAT Ibiza Mk1 that I used to go past doing my paper round as a kid in Walsall. One day the bonnet was open, revealing the Porsche logo – to be fair, the guy probably kept the bonnet open all the time to show it off – and it intrigued me. It gave SEAT desirability and credibility brownie points, on top of the sporty stylings.

In 1986, SEAT were bought out by VW. But VW don't really need SEAT and haven't known what to do with them. That confusion has now fed down to the consumer. Cupra, which was introduced as a high-performance sub-brand in the mid 1990s, became a standalone brand in 2018. Cupra are the premium wing of SEAT, offering the sportier designs and driving experiences. I like the Cupra stuff. Their first car as an independent brand, the Formentor, was good-looking and the high-performance version was decent. The Cupra Born – their little electric car, first produced in 2021 – was more fun, better to drive and featured edgier design than the equivalent VW model on which it's based, the ID.3. The trouble is, the Cupra stuff is basically the sportier, fun offerings that we really liked about SEAT, so we're left wondering: what's the point of a regular SEAT now? It's superfluous. They're like the appendix of the VW group.

Skoda

In every playground in the 1980s, someone would be cracking the gag:

'What do you call a convertible Skoda?'

'A skip.'

Or 'How do you double the value of a Skoda?'

'Put fuel in it.'

Skoda were a laughing stock. And when they turned things around, they took a gamble with their new advertising campaign. They chose to own their terrible reputation. There's an ad of a car transporter driver parking up outside a car dealership and pushing the button to lower the first car down on to the road. As the car comes down, the transporter driver looks more and more puzzled. Then he glances over to the big 'Skoda Auto' sign outside the dealership. He shakes his head, sighs and pushes the button again, loading the car back up. There's no way they can be Skodas – they look decent. He's picked up the wrong bloody shipment. To the Brits, who do specialise in being able to laugh at themselves, it was a masterstroke.

Even in their homeland of Czechoslovakia, where Skoda were founded in 1925, they had an image problem. Although they were named after their founder, Emil von Škoda, the word '*Skoda*' in the Czech language means 'pity' or 'shame'. That's a tough sell. Lada had a similar problem trying to flog their 'Nova' in Spain, seeing as '*No va*' in Spanish basically means 'Doesn't work'. That was unfortunate, but small fry compared to what the Hyundai Kona (or cona) means in Portuguese . . .

As a kid, the sight of a Skoda did make me feel both sick and sad, and that's not a great cocktail. The Rapid was an exception. Now, I think it looks cool, but I was too young at the time to appreciate the fact that it mimicked a Porsche 911 with its rear engine and rear-wheel drive. Yes, the top speed was about 90mph and I'm pretty sure you could complete a skydive in the time it took you to reach 60mph, but as a young man in the early 1990s, the Rapid was a decent car for drifting. It helped that it didn't look like a Skoda, because if your mates knew your parents owned one, it was a bit like putting your hand up in class and saying 'Mum' instead of 'Miss'. You ain't going to live that one down.

Things changed for Skoda when they entered into a joint-venture partnership with VW in 1991. In 1994, they brought out the Felicia, which was a reworked version of the Favorit, with a 1.3, 1.6 or 1.9 D VW engine. They'd produced a normal, dull, blend-into-the-background car. But that was a big step up. It's hard to imagine a corporate aspiration being 'looks inoffensive' but that's what they aimed for, and, in fairness, they nailed it!

When the new Octavia came out in 1996, Skoda really leant into the fact that parts of the car were from a VW Golf. And that appealed to people who both love a bargain and love to tell people, in extremely boring detail, about how they've landed themselves a bargain. They feel like they've won. They're in a secret society of like-minded geniuses, who will tell you proudly that they haven't followed the sheep and bought a VW. They've bought the same car for less money. Also, because their expectations were low to begin with, the cars performed much better than Skoda buyers imagined.

It's part of the reason why, when the results of owner satisfaction surveys were published, Skoda performed really well.

The upshot of all this was that Skoda started to build a following of advocates. Throughout the noughties, an army of self-congratulating Skoda buyers were watching their kids play football on a Saturday morning, rubbing their hands together in anticipation of the question from a fellow dad: 'What car are you driving?' And they'd start the answer: 'Well, actually . . .' with a smug smirk of self-satisfaction.

As a guy from a working-class town in the Midlands, I love a bargain. I not only like to know that I've cut a deal – I'll also broadcast it. So it may not surprise you to learn that I love Skoda. And, like a permanently smug Skoda owner, I'm going to tell you why. They're often slightly bigger than the VW equivalent, their pricing structure is clever and their marketing is brilliant. Modestly admitting that you're crap is such a win in Britain. Southampton were bottom of the Premier League at the time of writing and losing every game by a lot. They were being mauled by Chelsea the other day with eighty minutes on the clock and the Southampton fans started singing: 'How crap must you be, it's only 4–0.'

Skoda have begun to differentiate themselves from the competition by including lots of clever features in their cars. It's a sensible idea because the Skoda USP of value for money is being eroded by rival Korean and Chinese manufacturers, who are all producing cheap but solid cars. Car reliability is so good these days that every car manufacturer is now chucking features into their cars to try to offer something a little different. Many of these are flashy gimmicks, like the Mercedes dash cam so you can have a Google Hangout in your car. Who's going to use that? But Skoda have actually come up with some helpful features. They were the ones who thought up the sleek umbrellas that you slide out of a special compartment in the doors. Most people associate that feature with the Rolls-Royce Phantom, but Skoda introduced it a year before, in 2001, in the Octavia, Fabia and Citigo. Their 'Simply Clever' gadgets have made Skoda a talking point again, and this time for the right reasons. I've just checked and the umbrellas are quite pricey on eBay, so if you're in the market for a used Skoda, after you've kicked the tyres, make sure they've still got those funky umbrellas!

In the fuel filler cap of the new Octavia there's an ice scraper, and under the bonnet there's a little funnel built into the windscreen washer reservoir. These are genuinely useful. No one's got a funnel lying around unless your partner owns a Shewee. There are little grippers in the cup holders, so you can open a bottle one-handed and there's a net on the underside of the parcel shelf so your shopping doesn't slam into the sides of the boot. Some of the 'Simply Clever' features are a bit hit or miss. I'm not sure about the parking ticket holder or the magnifying glass on the side of the ice scraper, unless you're one of the five Skoda drivers still using an *A–Z*.

Skoda have come a long way. If you told me back in the early 1990s, when people were laughing about the Skoda Rapid, that ten years later Skoda would name a car the Superb, I'd have thought, *Yup, that sounds about right*. Another ludicrously ambitious name. I wouldn't have imagined for a second that it would actually be superb. But it is. It's huge and comfortable – sure, it's slightly boring, but if you're doing loads of miles on the motorway with the family, get yourself a fourth-generation diesel Superb. If you're a cab driver, you don't need me to tell you to get an Octavia. You're already driving one. Taxi drivers want to choose a car that won't break down and isn't expensive to run. There's a reason you don't see Uber drivers in a second-hand Jaguar.

BEST CARS OF ALL TIME
TO BE DRIVEN IN

1. **Rolls-Royce Phantom (either Mk1 or Mk2)** – The ultimate in motoring opulence, this car glides down the road and it's everything you ever imagined a Rolls-Royce should be.

2. **Toyota Century V12** – The only V12-engined car that came out of Japan. It's essentially Japan's Rolls-Royce, but being Japanese, it's far less ostentatious. It's also better built.

3. **Blower Bentley** – A car to be driven in rather than drive because it's quite confusing to operate, plus you don't want to damage it because they're worth about £20 million. But what an experience to be driven in a classic 1930s car that can go like a modern hot hatch.

4. **Lexus LM** – Basically a minivan that has been converted into a limousine with a luxurious interior, including a massive cinema screen.

5. **Rimac Nevera** – If you want to go on a fairground ride without having to actually visit a fairground, take a ride in this car. It launches like a rollercoaster and takes your breath away.

6. **Mercedes EQS** – An electric version of the S-Class and it just works because it's even more comfortable and luxurious and, because it's electric, it's even quieter than an S-Class.

7. **Skoda Superb 2.0 TDI** – A brilliant, practical and economical family car that's great for covering miles. It's well-made, good value for money and a terrific all-rounder.

8. **Range Rover Long Wheelbase** – If you want to be driven in the utmost luxury off-road, there is no better way to travel.

9. **BMW 7 Series (G70)** – A brilliant car to be driven in because it's super comfortable, very spacious and it has the coolest TV of any car. It's like being driven around in an IMAX cinema.

10. **Audi S8 (D5)** – If you're being chauffeured and are trying to escape some bad guys, what you need is an Audi S8. It's got the same engine as the RS6, yet it's disguised in an A8 limousine body with luxurious suspension and a cosseting interior.

Subaru

The trouble with Subaru is that any car that isn't an Impreza gets the 'What's that about?!' response from the public. It's a bit like when a nineties band does a 2020s comeback and has the balls to play new songs. 'Oi! Stick with the good tunes!'

The Subaru Impreza was a mega car. In the nineties, it was locked in a 'battle' with the Mitsubishi Evo and you were either Team Evo or Team Impreza. I was an Evo guy (more on that in the Mitsubishi chapter), but I did love the high-performance Subaru Impreza WRX. Back in October 1992, when Charles & Eddie were topping the chart with 'Would I Lie to You?', the first-generation Impreza WRX was rolling into view. Not that you'd be able to hear that song (or any song) when an WRX rolled into view.

It's hard to imagine that the first-generation regular front-wheel drive Impreza was basically a family car. The WRX wasn't. It was nuts. And it got more nuts. Each iteration of the WRX acquired an affectionate nickname based on the shape of the headlights. That kind of treatment is reserved for brands with a hardcore following, like Porsche with their fried egg head-lights on the 996. The second-generation WRX, produced from 2000–2004, was known as 'Bugeye'. Then, at the end of 2003, the 'Blobeye' came out. Then there was the 'Hawkeye', which debuted in 2006. Inside, almost all of the Impreza models were quite cheap and plasticky, but more importantly lightweight, which is good for what you want the car to do. And then there was this mad 'Boxer' flat-four engine. Combined with the unequal length headers (the first part of the exhaust system where the gases from the cylinders collect), you get this unmistakable throaty growl. With the off-beat sound it produces, it's like the John Bonham of engines.

During the tail end of the *Max Power* car-tuning scene in the 1990s, the Impreza WRX was a modder's dream. You could do anything you wanted to an Impreza both inside and out – customised bodywork, exhausts the size of tank muzzles, and, of course, massively tuned engines. With the WRX, Subaru offered something unusual to the British motoring public: it was an incredible car at an incredible price. We may not agree on everything in the UK but there's one thing that unites us. We bloody love a bargain. You could pick up a grey import (i.e., one imported unofficially not through the distributor) of the WRX STI in 1999 for under £25k. It was insane what you were getting for that price. There was no car faster down British country roads. Forget about Porsches or whatever else because it wouldn't matter – an Impreza WRX would absolutely destroy them. In a straight line or on a race track it wouldn't, but on an unfamil-iar section of road, the Impreza was unstoppable. The grip that car generated and the engine's ability to deploy its power so well was unreal. It meant that anyone with reasonable driving ability would skin a Porsche. It was almost unfair.

Unlike in a sports car, where you're sitting dead low to the ground, you're higher up in an Impreza, which fills the driver with confidence. The trouble is, combining confidence with a very fast, very capable car meant that it was all very good until it wasn't very good. My mum worked with deaf kids and that meant spending time in the company of ear, nose and throat (ENT) consultants. One of them was a guy who obviously had a bit of cash and he bought himself an Impreza WRX to play with. Of course, he hooned it around all over the place and had the time of his life until he crashed into a tree and broke his nose. In what sounds like a scene from *John Wick*, he politely knocked on someone's door so he could go to their bathroom and reset his own nose.

The WRX STI was the even higher-performance version and it was a special car. In the right hands. My mate, who I used to work with on *Auto Express*, used to be a Subaru car salesman. He told me about this old boy who'd been a Subaru customer for ages and always bought the top-of-the-range car. The Impreza STI was the most expensive model, so he bought that, not really understanding what it was. And to be fair, if you're just driving it to Waitrose, it's going to feel like a normal car because the turbo only kicks in above 3,000rpm. So this old boy has no idea that he's sitting on a rocket ship of a car until one day he has to head out of town. He gets on the motorway so, of course, the revs go over 3,000rpm, the turbo kicks in and it frightens the shit out of him. He got off at the next junction, drove at about 25mph for the rest of the journey, and returned it to the garage before waddling home in what I imagine were soiled trousers.

The specialised wide-bodied, limited-edition STI 22B was legendary. Subaru produced 399 of them for the Japanese market and they sold quicker than a dog barks when it hears the exhaust. They built sixteen of them for the UK market and three development cars, one of which was sold to rally GOAT Colin McRae. That car sold for £480k in 2024. In 2022, the motorsport engineering group Prodrive produced a restomod (a restoration combining the design of an older car with modern tech) based on the shell of a two-door STI that they'd rebuilt – with a lot of carbon-fibre – and debuted it at the Goodwood Festival of Speed that year. It was called the P25 and I did enjoy the numberplate on the press car: P25 POA, reflecting the car's starting price of £525,000. The fact that they sold the full run of twenty-five cars in advance tells you a lot about the special-edition Impreza fan base. It's hardcore.

I know some of you are now going to flick past this next bit, but I have to say it: Subaru have made some other impressive cars. The Outback and Outback Sport (both estates, produced from 1994) were cars that were bought by people who knew about cars. The first-generation Forester (a kind of tall, wagon version of the Impreza, first produced in 1997), with its 2.0-litre WRX engine, was a good car and fun to drive. Toyota have owned Subaru since 2005 and they jointly developed a car in the early 2010s. Some arrangements are more equal than others, though, and the car was manufactured at Subaru's plant and featured a Boxer engine, which is a Subaru thing. The Subaru model was called the BRZ and the Toyota was the GT86 and the Toyota massively outsold the Subaru. Part of that was because Toyota tweaked the suspension to make it more fun to drive and sportier. Meanwhile, Toyota told Subaru to build the car and then how to position it in the market, while making sure that their GT86 was marketed and positioned better. It feels a little bit like Subaru got shafted, essentially.

They are also guilty of self-shafting, however. There was a period when Subaru nailed the shape of their cars in the late 1990s, but then they conformed to the age-old approach of changing the design to get people to buy the next one. To be fair all the manufacturers do it; it's just that some nail it, like Peugeot did when the first generation 3008 metamorphosed from dull-looking caterpillar to butterfly with their impressive second-gen offering. Some manufacturers, however, take something that's already excellent, like their late nineties range, and then absolutely screw it up. If you look at Subaru's range now, I don't even know what it is.

It doesn't help that Subaru don't have manufacturer-owned

UK operation. Instead, in the UK, Subaru cars are sold via a separate company called the IM Group, which is basically a distributor that does the PR, so they don't have the budget to push the cars as they would if Subaru were properly based here. But maybe they don't need to be. We're still all talking about Imprezas either way.

Suzuki

I've got a soft spot for Suzuki because they make the Jimny, one of which is proudly parked in my driveway. It's a funny little car and, to be fair, Suzuki's a funny little company, in the UK at least.

There's very little brand loyalty when it comes to cars any more but Suzuki buyers are different. When I was a kid, it was either Ford or Vauxhall. Aside from a couple of experiments with other brands, my dad always bought Fords. Never a Vauxhall. Meanwhile, my dad's good mate only bought Vauxhalls and never a Ford. Suzuki buyers seem to be similar. They're a bit like the folks who buy their clothes at Tesco. It's not modern and it's not flash, but it's functional. They know what they're getting, they haven't been let down, so they stick with them. There's a certain type of bloke working at Suzuki. Let's call him Brian. Brian's not trying to smash any sales targets. Brian's holding the fort. Likeable and dependable, Brian would probably invite his customers to his sixtieth birthday party, and I'd wager that a fair few would attend. Every couple of years, Brian calls up his customers, probably from a landline, and they answer, also on a landline.

'We've got the new Swift for you, Mr Watson.'

'Lovely. I'll pick it up on Tuesday after popping into Tesco.'

What's unusual about Suzuki is that you'd expect them to be consistent across the different products they make. But in the superbike world, Suzukis are bloody mental. I wonder if the guys who work on the superbikes are a bit ashamed of their car arm; like they're the cool older brother and their younger brother's a bit of a dweeb by comparison. That said, their cars are generally pretty good. There's something honest about them. You don't feel like you're paying extra for a shiny showroom. The Suzuki customer doesn't want that. They just want a decent car that gets them from A to B and doesn't cost a lot to fix. It does the job. But what you get with a Suzuki, perhaps slightly unexpectedly, is a bit of character. The Swift has a light, agile feel about it. It's not fast, the engine's a bit coarse, but it's got something. And that's why I think their owners like them, because they know that the next Suzuki that Brian tells them about will, like Brian himself, be honest and trustworthy. The Swift is an unsung legend of the car world, so it's time to start singing. Suzuki have sold nine million of them. It's been produced for over forty years. It's time to get to know your local Brian.

When you build up a relationship with a car manufacturer, you give them a bit of a pass on some of their weaker points. I thought the Suzuki Jimny was cool immediately, but don't expect it to behave like an SUV. For one, it's terrible on the motorway and you're constantly having to put steering corrections in. The back seats are tiny, there's no boot in it, it's slow, and it weaves about the place. It's great off-road, not that anyone's actually going to use it off-road. But the one thing it will do is make you smile. When I reviewed the Jimny, I gave it a six out of ten. My feeling was: it's probably not for you. And then I went out and bought one. I'm lucky. If I only had one car, I wouldn't buy a Jimny. But because I have access to other cars and I like cars, I respond well to a car with character. It feels like a car for people who like cars.

I've currently got four Jimnys on a boat from Japan, because you can't get the passenger car version of the Jimny in Europe (you can only get the van version) because the emissions are too high. Incredibly, emissions-wise, its 100hp 1.5-litre engine is not far off my Porsche 911. So Suzuki stopped selling Jimnys because they were facing fines from the EU for going over their emissions quota. Japan drives on the right, so I won't need to sort that out – I'll just need to do the paperwork and have a rear fog light added so it'll pass the MOT. I've imported them for a video because their cult following means that even with tax import costs and delivery factored in, I should be able to sell the Jimnys for a profit.

Suzuki Jimnys hold their value ridiculously well. I bought my Jimny in 2021 for something like twenty-one grand. It's got about 8,000 miles on the clock and I reckon I'd probably get twenty-six grand for it now. It's appeared on a few Carwow videos, like the one in late 2024 where I drag-raced it against a Hummer. The idea was to pitch two off-roaders against each other, one small and light, one big and heavy. The Jimny took the Hummer down over the quarter-mile! And it won the brake test as well because, well, I think we're still waiting for the Hummer to actually stop.

I just drive my Jimny a little bit around the village and not much further. Everything that can be added to this car, I've done, to make it look cooler. It's got a diff guard underneath, side strips, a shiny wheel cover at the back, a tow bar and a roof rack. I've never towed anything with it or used the roof rack. I've just realised that it sounds like I'm writing an ad on Carwow for my Jimny but it's not for sale. And that's for one very good reason: when my daughter was born, I picked her up from hospital in it. That means it's staying with me for ever.

Tesla

Before the first Tesla model rolled off the production line/was parachuted down from near Earth orbit, all the motoring journalists had the same question: is this genuinely *giant leap for mankind* stuff, or are we going to find a bloke behind a curtain?

You have to give Elon and co credit, though, because, like *Star Wars* or *Back to the Future*, even the opening credits have everyone whispering to each other excitedly. You only have to walk up to the Tesla Model X and already these DeLorean-like falcon doors are raising their wings up, like a sort of intergalactic salutation. Then you just sort of float into your seat. After that, you can't help yourself from opening and closing the magic doors. Again and again. It's just one of many features that have you grinning and giggling like a child. Like when you're pootling along at 30mph, and then you put your foot down and you're suddenly at 70. That instant zip from the electric motors means you accelerate just for the hell of it. Again, much like a child, I found myself enjoying having my head yanked backwards and forwards. It's ridiculously addictive. And that's just in the Model X. When you're in the Model S Plaid, even the Bugatti Chiron Super Sport is going to be in your wing mirror. Because you'll be doing 0–60 in 1.99 seconds. Just to put this in perspective, the Bugatti Chiron Super Sport is an 8.0-litre quad-turbocharged two-seater supercar generating 1,600hp. The Tesla Model S is a family saloon.

One thing that people don't tend to know about until they've driven a Tesla is that there's a whoopee cushion feature. That's right, the driver can select a fart sound to time perfectly with one of your passengers taking their seat. But there isn't just one fart noise; there's a whole range, from 'short shorts ripper' to the worryingly wet-sounding 'neurostink'. These cars are built for adults who have never grown up. And to be fair, that's almost all of us.

Even counting other Tesla models, the Cybertruck is the most hyped-up car I've ever known in my twenty-five years as a motoring journalist. It completely divides opinion in the motoring world but I think everyone can agree that it's completely bonkers and no one except Elon Musk and Tesla would attempt it. But then you see it, and it just looks amazing. There's nothing else like it. The PR soundbites for the Cybertruck start normally: it's billed as the fastest electric pickup truck, and then they get increasingly mad. Namely, its steering wheel isn't connected in any way to the wheels and it's bulletproof, exactly what I'm looking for in the mean streets of the Cotswolds.

The value-for-money devil on my shoulder starts whispering: *Why build it out of this stainless steel stuff? It's hard to produce and what's it going to be like in terms of maintenance?* But then I look at it again and that metallic finish makes every negative thought leave my brain. Is Elon pumping something out through the air vents, I wonder?

Tesla invited Carwow, as one of a small selection of content creators, to film early exclusive content on the Cybertruck at their Gigafactory in Texas. It was an amazing experience driving the brand new high-performance version of the Cybertruck – the Cyberbeast – and drag-racing it on an access road outside the factory. For one of the drag races, we pitched the Cyberbeast against a Lamborghini Urus, a Ford Raptor FR 50 and a Hummer Lightning (the fastest electric truck before the Cybertruck). Getting the opportunity to do this stuff is super cool.

Naturally, we thought of some silly things to do with the

Cybertruck. You can't help it when you find out that some-thing's bulletproof. You want to find out how bulletproof. But we're British so, rather than fly in a chopper with a chain gun, we wheeled a shopping trolley into it. Then we all kicked it a lot. In a first for me, I did ask the guys at Tesla beforehand: 'Is it all right if I kick the crap out of it?' 'Knock yourself out' was the reply, which was possibly delivered a little seriously for my liking. We had a great time filming it, but annoyingly, when I released the video on my personal Twitter account, someone ripped it from there and posted it to their own account. Sometimes that happens, but this time, Elon Musk only went and retweeted that guy, so he ended up nabbing all my views and likes!

The Cybertruck does look incredible, and it drives really well, especially when you consider that it weighs three tonnes. It's got very clever adaptive suspension and the steering adjusts to the speed you're driving at, making this three-tonne monster more manoeuvrable than you could ever imagine. It feels much more like an SUV than a pickup. And that's part of the problem: it's not going to be used as a pickup. It's got road presence, it's spacious inside, it fits your lifestyle if you want to stick bikes in the back but it just doesn't feel like a working vehicle.

The other snag with the Cybertruck is it's so unique that when more than one appears in front of you, it's a bit like seeing how a magician's performed a trick. It's still cool but just a bit less cool, if you see what I mean. If you've bought a Cybertruck, you don't want to be parking it next to another one, really. It feels like less of a statement and more of a motorcade. Also, I'm just not sure about it in the long term. It's more expensive than ori-ginally planned and the range has been dialled back. And how you clean the inside front corner of the windscreen is anyone's guess. Maybe there's a satellite-controlled cleaning robot. The Cybertruck just feels a little bit like hubris, which is in keeping with the man behind it. There have also been lots of complaints about build quality, about its actual usefulness as a truck, and a potentially dangerous bonnet closing issue. The edges are sharp at the front as well so it's never going to pass any UK or European crash regulations.

So how many have they sold? Tesla don't publish the figures but, fortunately, when you issue as many recalls as Tesla have, you do get a window into sales. In the eighth recall – this one for a defective adhesive that could cause a stainless-steel trim panel to fly off the car – all Cybertrucks built since 13 November 2023 were recalled. That's 46,096 of them – some way short of the figures that Tesla had been putting out. Over 1 million people reserved one. And then Elon made a song and dance about being able to produce 200,000 a year. He's not going to need to.

BEST GAME-CHANGER CARS

1. **Ford Model T** – An iconic creation that brought cars to the masses thanks to clever production techniques that are still used today.

2. **Tesla Model S** – Until the Model S came along, electric cars were budget-y and boring, but this one had decent range and a luxurious feel to it. It made an EV desirable.

3. **VW Beetle** – Designed by Ferdinand Porsche (although the less said about who commissioned it the better), the Beetle was meant to get the people of Germany moving. It got the world moving. Production of the Beetle only reached the end of the road in 2019.

4. **Mini (Mk1)** – The brainchild of incredible designer Sir Alec Issigonis, the original Mini was brilliant to drive, cheap to own and surprisingly practical.

5. **Porsche 911** – Redefined what a sports car could be when it was launched in 1963 and it's still the benchmark of how a performance car should be. Rear seats add practicality, which make it the only sports car that you can buy with your head as well as your heart. Whenever motoring journalists grade a sports car, they always compare it to a 911. The 911 does it all.

6. **Toyota Prius** – Mating an electric motor to a petrol engine meant that you could recoup power under braking and you could drive the car for small periods of time on electric power alone. It gave you diesel economy but without emissions. While the original version looked crap, the second generation looked great and continued

to deliver impressive economy. There's a very good reason that when you step into an Uber, the chances are it's a Prius.

7. **VW Golf GTI** – It wasn't the first hot hatch, but it was the first usable everyday example that made hot hatches popular, one of the reasons being that it was fitted with fuel injection, which meant it worked every time you turned it on.

8. **Audi Quattro** – A car that reshaped rallying but also showed that you could use a four-wheel drive system to give a car performance in all conditions, not only in motorsport but also everyday motoring. It changed the whole car industry.

9. **Ford Mustang** – The original Mustang blended performance with style and affordability. It created the 'pony car' segment of the automotive world, and many other manufacturers followed suit. It meant that you could have a practical, fun and good-looking fast car at a decent price.

10. **Nissan Qashqai** – It wasn't the first crossover SUV, but it was the first one that was bought en masse. It drove well, was very practical, looked good and it gave people that sense they were in a 4x4 even when they weren't. It was the car that turned Nissan's fortunes around and is the reason why everyone's driving an SUV. Blame the Qashqai!

Toyota

When I was seventeen, I wrote off my dad's horrible Metro, so I ended up using my mum's H-reg Ford Fiesta XR2i. Compared to the Metro, that was a bloody supercar. One day I was with my mate Greg, who was always talking about how fast hot hatches were, and we ran into his friend who had a Toyota MR2.

'That's way faster than your car, Mat,' he said.

Game on. So we had a little street race, as you did back in the day, when you're young and foolish and don't realise the potential consequences. Only it wasn't so much of a race as it was me ingesting the exhaust fumes of a superior car. That was the moment I went from being a guy who liked the look of cars and the idea of cars to someone who started to geek out about the mechanics of cars. I'd be lying if I told you I wasn't motivated by wanting to understand why I was eating someone else's dust. Both the XR2i and the Toyota MR2 had a 1.6-litre fuel-injected engine, so it couldn't be anything to do with that. But the MR2 did have way more horsepower. It's also rear-wheel drive with a mid-mounted engine, so the engine was in the right place for putting traction down and turning, whereas the XR2i is based on a basic little front-wheel drive super-mini, just with a slightly higher performance engine inside. The MR2 was a newly built, aluminium-headed, dual-cam, 16-valve, high-revving, 122bhp engine. When I got in the MR2 and gave it some gas, it just felt like it wanted to show what it could do. I got back into my mum's XR2i. It was so sluggish and lethargic. Christ, that was a boring journey home. But that day I realised how different two cars could be, despite, on paper, having the same kind of engine.

The MR2 was suddenly on my radar, and that was the gateway into Japanese cars for me. At that point, in the early 1990s, Toyotas were so cool. The fourth-generation Supra was one example. It was a twin-turbo powerhouse, generating 326bhp, and had a leather interior, traction control, ABS – it even had heated mirrors. Only about 600 of them were sold in the UK between 1993 and 1996. After that, people started importing the Japanese versions and I'd be on the lookout for them. The sixth-generation Celica came out in 1994 and I loved that as well. My family, on the other hand, weren't team Toyota. They'd be talking about Fords and Vauxhalls.

When I became a motoring journalist in the early 2000s, Toyota were transitioning very successfully from really cool to incredibly dull. Aside from the Prius, which was quite a ground-breaking car, they focused on normal, sensible, reliable cars. It continued this way into the 2010s. I remember shooting a video in 2015 of a BMW M4, and the intro involved parking the car next to a small, boring Toyota hatchback. Then I said to camera that I was going to review a very exciting car. Then I got into the Toyota, shuffled across into the passenger seat and climbed into the BMW without mentioning the Toyota. That glib intro didn't win me any friends at Toyota UK. But that has long since changed. As has their car line-up.

The current chairman of Toyota, Akio Toyoda, had his work cut out when he was appointed the president of TMC (Toyota Motor Corporation) in June 2009. Toyota were trying to ditch the dull tag while steering their way out of the global financial crisis, which had already claimed General Motors and Chrysler. It wasn't about to get any easier for Toyota (and Toyoda) because five months later, they were recalling over 5 million vehicles as a result of fatal crashes involving accelerator pedals trapped by ill-fitting driver's side floor mats. It didn't stop there. Two months later, they were recalling another 2.3 million cars with faulty accelerator pedals. And a month later, they issued a recall on the 2010 Prius because of an ABS 'software glitch'. They say bad news comes in threes, but bad news is something of an understatement. This was an absolute shitshow.

Many chairmen would have fallen apart, but Toyoda wasn't a typical chairman. As a young exec, he spent a lot of time on the front lines of the factory floor rather than the boardroom. Toyoda wanted to see what wasn't working with his own eyes, and what he witnessed was that production and sales weren't talking to each other at all. So he completely transformed TMC's domestic dealership network. After he became president, he made big changes, separating the main corporation into five companies (Lexus, Compact, Mid-size, Commercial/Vans and Gazoo Racing) and reducing the number of executives and senior advisers (all of whom were north of sixty years old) from 146 to fourteen.

But the thing that really annoyed Toyoda was the brand's association with dullness. This was a guy who liked a fast car, and Toyota weren't making any of them. In 2007, he'd set up TMC's Gazoo Racing to encourage employees who also loved fast cars to help create great Toyota sports cars in the future. At this point I should mention that Toyoda didn't just like sports cars – he was also an amateur racing driver. Although it's a great sign that the chairman of the company is into cars and races them, it did cause some safety concerns at TMC. So he came up with an alter ego – Morizo – and kept racing.

Toyoda admitted that two events in 2011 lit a fire under him. That year, Toyota unveiled the Lexus GS in Pebble Beach, California, only for an American journalist to comment: 'Lexus is boring.' The second event was at the Nürburgring, where Toyota was conducting test drives along with their rivals from BMW, Mercedes and Porsche. In what sounds a bit like the Jamaican bobsleigh team in the film *Cool Runnings*, the established European teams laughed at Toyota's offering. The shame that Toyoda felt was a big motivator.

Toyoda launched the Toyota GT-86 (a collaboration with Subaru) at the Tokyo Motor Show in December 2011. I tested it in 2012 and it was seriously fun – the driving position was perfect, the chassis was really responsive and the gear shift was awesome. You felt really immersed in the mechanics of the car. Toyota was back! That success was one of the reasons that Akio Toyoda won *Autocar*'s Man of the Year award for 2012, beating Land Rover's Gerry McGovern.

'Morizo' had the last laugh at the Nürburgring in 2014, competing at the 24 Hours race and finishing first in class (and thirteenth overall) in a Lexus LFA. In 2016, he championed the design and development of a new Toyota sports car (no collaboration this time), and this became the GR Yaris. Can you guess who the test driver was during the development stages? That's right – our mystery man Morizo.

When Toyota announced the GR Yaris, my first thought was: *You're doing what?!* It had a bespoke engine, a bespoke four-wheel-drive system, a bespoke body, a carbon-fibre roof and a limited-slip diff (LSD) on the back and the front axles. Porsche 911s don't have diffs on their rear axle as standard. McLaren didn't put limited slip diffs on their cars until the 750S in 2023. GR Yaris prices started at about £33k for the ones with two LSDs. If one of the German car manufacturers had made a car that bespoke, it would have been double the price. I'm convinced that they must have made a loss on that car, or did some clever accounting by putting a lot of the development budget through the Gazoo

Racing Team. But what a car that turned out to be. Toyoda was living up to his word. No more boring cars.

Sometimes things get a little bit lost in translation when communicating remotely with Toyota in Japan, but when you actually go there, meet them in person and start bonding over a few sakes, all that magically disappears. They're amazing hosts and really proud of their culture, and seeing you take pleasure in that establishes trust between you. And once that trust is in place, the access you're granted at Toyota is quite unlike any other car manufacturer.

I found myself at a completely mad bespoke event with cars that I'd never driven before. I was the first journalist in the world to be shown around their new GR factory. I came up with an idea over lunch to bring in a racing driver on a track and time the different models of GR Yaris. 'We'll see what we can do,' was the answer. If a Brit says that to you, it's a hard no. In Japan, it means 'We will stop at nothing to make this happen.' The next day, they'd found a track and a racing driver and transported over a Mk1 and both manual and auto Mk2 GR Yaris. Just so I could shoot a crazy-ass video. This is just one of the reasons why Japan is my favourite country. I love the people, I love the food and I love the cars.

To be fair, as soon as I saw the GR Yaris, I wanted one. Then when I finally drove one during the launch in the UK, after driving it just 100 metres up the road I knew the car felt right and that I'd own one. In fact, I loved the car so much that I bought the Mk1 I'd been using as a long-term demo from Toyota.

When I was over in Japan at the launch of the second-gen GR Yaris, I found myself talking to the chief engineer of their GR division. I'd never met him before but he came over, shook my hand warmly, thanked me for our videos of the GR Yaris and told me how they'd helped show people how good the car was. One of the GR engineers there even showed me the internal documentation that Toyota used to develop the second-gen GR Yaris, which included one of my videos where I was talking about an issue I'd found with the car. They used my video as a reference and fixed the problem I'd identified in the later model. Toyota do things differently.

Case in point: some car manufacturers tell me off for sometimes doing this thing where I throw the parcel shelf if it doesn't fit underneath the boot floor. At Toyota, though, a group of the engineers actually asked me to throw the GR Yaris's parcel shelf because apparently they love it when I do that. They even gave me a round of applause while I was doing it, which was kind of crazy in an amazing way. I loved my Mk1 but the Mk2 was even better, so I upgraded.

I've got a habit of importing cars from Japan at the moment. One of them, which I'd been thinking about buying for a long time, is a Toyota Century. The Century is a four-door limousine with a naturally aspirated V12 engine – the only Japanese car with a V12 – that the Emperor of Japan used (with a few mods) as his official state car. And that's fitting – the Emperor is treated like a holy figure in Japan, and so is the Century. The interior and exterior of the car, hand-built by master craftsmen, is symbolic of Japanese values. Their marketing slogan is 'The Century is acquired through persistent work, the kind that is done in a plain but formal suit.' It's all about understated elegance, with clever touches that only the Japanese would have thought of. The paint finish is so good and is polished so rigorously that the rear panel by the back window is designed to act like a mirror so you can check your appearance before you go to your meeting or step on to the red carpet. The interior fabric is wool, not leather, because moving across leather upholstery can produce a squeaky sound, which is considered both cheap and socially embarrassing. Like the country it comes from, the Century is unique and special.

And on that note, I can't not mention the Toyota Land Cruiser, which is, incidentally, also used as the official state car in several countries. To shoot a series of videos, Carwow bought a 200-series Land Cruiser – it had 280,000 miles on the clock, which is considered just about run in. That V8 in the Land Cruiser is one of the most reliable engines in the world. They say that if you want a car that can take you into the wilderness, a Land Rover is a good option. But if you want to get back out of it, you need a Land Cruiser.

BEST CARS OF ALL TIME TO DRIVE ON A COUNTRY ROAD

1. **Porsche 911 S/T** – Probably the best 911 ever built. It has an old-school analogue feel and race car purity but in a package that's perfect for the road.

2. **BMW E46 M3 CSL** – BMW somehow removed 110kg from the M3's body and gave it more power to create one of the most finely balanced driver's cars of all time. Back when it was released, motoring journalists said it wasn't worth the extra money but oh how wrong they were!

3. **Ferrari 458 Speciale** – The ultimate form of the 458 with an amazing, naturally aspirated V8 engine – the last version of that engine before Ferrari went all turbocharged.

4. **McLaren 675LT** – Has a bonkers engine, a brilliant chassis and the best steering on any road car I've ever driven. It's a pure delight.

5. **Toyota GR Yaris Mk2** – You can probably drive this faster down an unfamiliar road than any other car. And you'll have an awesome time while you're doing it.

6. **Mazda MX-5 RF** – Gives you the best of both worlds: it's a hardtop but you can also take the roof down. The brilliant lightweight chassis of the MX-5 is so fun and playful.

7. **Lotus Elise Mk 1** – The lightweight aluminium chassis, zingy Rover K-series engine and the ability to take

the roof panel out guarantees smiles for miles (until something breaks on the car).

8. **Alpine A110** – Basically a French version of a Lotus Elise but for the modern day. It has a punchy turbocharged engine and a terrific lightweight chassis and suspension that glides over the road much better than you imagined a sports car would.

9. **Mitsubishi Evo VI Tommi Mäkinen** – Made to celebrate Tommi Mäkinen's World Rally Championship wins, this is a fantastic rally-bred saloon car with incredible electronics that add to the driving experience; it's the old-school version of the Toyota GR Yaris.

10. **Caterham Seven 310S** – Its 1.6-litre engine has plenty of punch in a lightweight body. Being so exposed to the elements and seeing the suspension working and the wheels moving makes you feel like you're in a racing car.

US Manufacturers

One of my favourite driving experiences ever was in a Jeep in the States. I flew over with Carwow to Los Angeles to film a lot of American cars and we'd teamed up with edmunds.com (a big car-shopping site in the US) and got to film some of the cars that they had in their fleet. One of the cars was a Jeep Gladiator pickup, with a suspension lift kit, massive tyres and the body panels taken off. We filmed the Gladiator driving through canyon roads and on through the hills about ninety minutes from Venice Beach while the sun was setting. The air was warm and there was this lovely growl from the V6 engine going on. The car wasn't particularly fast and didn't handle very well, but none of that mattered because you couldn't help but be in awe of the American showcase surrounding you. Huge highways, huge car, vast landscape and equally vast burger, when we stopped to grab some food. Honestly, it was one of those sensations that isn't necessarily about the dynamics of the car – it's more about the feeling the car gives you in that given situation. The video did pretty poorly, but it was one of my enduring memories of driving a car. I dreamt of long, beautiful American road trips and I've been lucky enough to do quite a few. I did one in a V8 convertible Ford Mustang driving along the Pacific Coast Highway in California, stopping off at beautiful beaches and looking at the sea lions. But that one in the Jeep topped them all. It was only two hours, and yet it's etched itself into my memory.

Jeeps are very good at what they do. I've taken many of them off-road and they perform really, really well. They have loads of quality off-road tech on them, like you can decouple the anti-roll bars so the wheels can move more freely when you take the car off-road. They have diff locks, front and rear, which makes the axles spin at the same speed, increasing traction. *On* the road, a Jeep is a bit like a big Suzuki Jimny – i.e., it's pretty rubbish in lots of ways – but it's got character and that counts for a lot.

The brand image of Jeep is very clear, which makes the European spin-offs they do really weird. Yes, the Jeep Avenger (a crossover SUV and the smallest car Jeep make) won European Car of the Year in 2023, but I'll refer you to my feelings about that particular award on page 209. In my eyes, the Avenger is in no way, shape or form a Car of the Year winner. It's a bit expensive, very middling in terms of all-round performance and features, but it does look good so I could see why people would like it. It looks a bit Jeep-y but it's really not Jeep-y once you try it.

Jeep feels like an outlier in its parent Stellantis group, but then again Stellantis does own Dodge as well. They were both part of Fiat Chrysler, which merged with the PSA Group (Peugeot, Citroën and Co.) in 2021 to form Stellantis, so both brands have been bounced around a bit. But Jeep in particular is such an iconic American brand that being part of this European mixed bag of a company feels wrong. I'm surprised Donald Trump hasn't reclaimed it.

The Dodge Ram is a very good pickup. I've driven them in America and they're popular, offer good value for money, and they feature big, reliable engines. Then there are the Dodge muscle cars, like the Challenger Hellcat Redeye, with its 6.2-litre supercharged V8 putting out 808hp and 959Nm of torque. I remember drag-racing one against a Porsche 911 Turbo S and it got absolutely hammered. Yes, the Dodge probably can do decent numbers if you have it on a prepped surface and everything's right at that point in time. But it does feel a bit all fart and no poo. It's still fast, but Porsche make every single horsepower count, whereas the Dodge does seem to waste its horsepower on wheel spin and noise.

But there's still something slightly cool about them with the looks and all that fury. The problem with these American cars is that when you drive them in America, you're all 'Yee-ha!' and you get it – the massive roads, the huge ranches, it feels totally right. Although ironically the speed limits are mystifying in parts of the States. An arrow-straight highway that disappears into infinity, no one on the road, no visible signs of life, middle of nowhere, and yet there's a sign up telling you there's a 40mph limit. Maybe Wile E. Coyote's the one putting them up, trying to pull a fast one on Roadrunner.

In the States, where everything is vast, a growling muscle car works. Then you see one in the UK stuck on the North Circular in London and it's a bit like seeing a grizzly bear in a zoo in Devon. It's not where it's supposed to be. They're good value in the States, but once you've imported one to the UK, factoring in import taxes and all that kind of shit, you're paying the same money as a Porsche 911. The interiors in American muscle cars are generally quite plasticky and nowhere near as boutiquey and bespoke as European high-end manufacturers. They feel like they're built to do a job, and in fairness, they do that job. They're bloody reliable because the Americans can build engines. The American approach is unique compared to, say, the Germans, who seem to want to make very complicated engines that are smaller but somehow ramp up the horsepower. The American approach and answer to everything is *Why don't we just make it bigger.* Yet it does work in the States with their solid, reliable and great-sounding muscle cars.

A case in point is the Chevrolet Corvette Stingray and its

high-power version, the Z06, featuring the most powerful naturally aspirated production V8 engine. It's a mid-engined sports car and it's fast, it handles well; it's a really good car. When I launched one, though, it's very hard to get the numbers that are claimed. In a Porsche, the numbers they give you will be a worst-case scenario, going uphill on a slippery day in a car driven by the engineer's grandmother. Whereas Americans, the kings of marketing, will have achieved the numbers they claim once in a cyclone of a tailwind. Would I have a Stingray over a 911? No. But if I lived in the States, sure, I'd be interested, because over there they're way better value.

Another cool American car I've driven is the Hennessey Venom F5 hypercar. John Hennessey, the founder and CEO of Hennessey Special Vehicles, made his name as a tuner but produced his own bespoke hypercar in 2020. He made no secret of his aim to make the world's fastest production car and his Venom F5 features a 6.6-litre, twin-turbo V8 engine named 'Fury' that produces an insane 1,817hp. In 2023, I got to drive it for a bit and it was crazy quick but John himself drove it for the launch we filmed. In July 2024, John set a new half-mile speed record in a Venom coupé, reaching 219.07mph and his test driver, David Donohue, took it even further, up to 221.92mph. In April 2025, Hennessey released the Venom F5 'Evolution', which drives up the power even more, up to a ridiculous 2,031hp. That makes it the most powerful internal combustion engine in a road car.

In 2023, I drove a Hennessey Mammoth, which is based on the Dodge Ram TRX pickup truck and then basically fed steroids to send it up to 1,026hp and an unholy 1,314Nm. The sound that truck makes is absolutely nuts. It's brilliant. It was a little sketchy hurling this beast around the narrow, windy country roads in the Cotswolds, where any vehicle coming the other way comfortably in their lane is a hazard. Even with the big old nobbly tyres,

I did 0–60 in 3.9 seconds and the quarter-mile in 12.34, which is beyond comprehension in a pickup truck. It took about fifteen minutes to fill up – not great when you've got a guy behind in a Vauxhall going quietly and then less quietly mental. He was wedged in between my Hennessey Mammoth and a white Transit van, who probably could have moved but was enjoying the drama unfold. One penny short of £185 went into that oil station of a fuel tank, and you need every fluid ounce of that when you're doing 8.5 miles to the gallon.

I have driven every generation of Dodge Viper, which was cool because that was a car I dreamt of driving. It was a car we knew in the UK because it was one of the fastest cars in the world and it featured in video games like *Forza Horizon*. It's a great car, with that mad 8.0-litre V10 engine. I remember the Mercedes UK press office buying a Dodge Viper from the States, and the story goes that one motoring journalist drove it around London and came to a width restrictor. Rather than just backing up or going through very slowly, he accelerated really hard to go through it, with his rationale (in his tiny little mind) being: *If I'm going quicker, I'm in the width restrictor for a shorter period of time, so less damage can be done.* He ripped the sides off the car. It was off the road for six months because they had to get replacement parts shipped in from the States.

Vauxhall

In case you're wondering why, when you go on holiday in Europe, all the Vauxhalls have a better badge and name, it's because Vauxhall and Opel are owned by the same company, Stellantis, and Vauxhall and Opel have been producing pretty much identical cars since 1980. Opels are mostly made near Frankfurt in Germany. Vauxhalls are made in Luton. This was one of the reasons why, when I was at *Auto Express* in the noughties, I wrote a column about why Vauxhall should give up the name, which I felt was holding them back, and embrace Opel, which is far cooler. The Vauxhall griffin badge has always been a bit weird but the Opel one, with its horizontal lightning bolt crossing the circle, is way better. It didn't help that back then Vauxhall was a crap part of London so a lot of people made that association. The area has since become cool and expensive. The cars have not.

In the 1980s in Walsall, it was all Fords and Vauxhalls. Most of my family were Ford folk, but my uncle was a proud Vauxhallian (let's call that a word). It felt like Cavaliers, Astras and Novas were the default cars around my part of the world. My girlfriend had a Nova and it had that unique combination of both being really loud inside but also really slow. But it took us everywhere. If you were really lucky, you'd see a Senator (produced 1978–1993), Vauxhall's executive saloon that was also available as a fastback coupé. Or an Omega, the sort-of successor to the Senator, made from 1986 to 2003. The posh one.

Part of the problem was that Vauxhalls were so mass-produced that you'd see them everywhere and yet you wouldn't notice them. 'It's just another Vauxhall,' you'd say, like you'd seen a pigeon. But if you saw one in isolation for the first time (a Vauxhall or a pigeon), you'd probably have a different opinion. The other issue was that they didn't seem to be owned by anyone who liked cars, and that was a problem. So you didn't really want to be the person with a Vauxhall. I can't help but associate Vauxhalls with a travelling salesman carrying around carpet samples in the boot. There's a reason that the producers of *The Office* stuck David Brent in a Vauxhall Vectra after he's been made redundant and he's tragically flogging cleaning products door to door.

The new Astra (the eighth-gen, produced from 2021) looks decent and it's decent to drive but no one gives a monkeys. People still get a little bit excited when Ford build something because there's the hope that it might be another groundbreaking car for its time, like the Mk1 Focus or the Mondeo, but there's no such glimmer when Vauxhall launch a car. I think that's the crucial difference between these two similar car manufacturers. Vauxhall haven't produced a mainstream model that was truly groundbreaking. They've had some sparkling moments, but when you look a little closer, some of these are basking in the reflected glory of another manufacturer. The Lotus Carlton was at one stage a Vauxhall Carlton, but Lotus went full Frankenstein's monster on it and turned it into a beast that bore little resemblance to its humble platform. It's someone else's car with a Vauxhall badge. The VX220 was also made by Lotus. The Astra GTE came out in 1983 but it wasn't groundbreaking – the Golf GTI, which had been released in 1976, took care of that. Vauxhall are a bit like Asda. Certain things they do are pretty good and they can occasionally surprise you. I genuinely like the George clothing range, but if there's a Sainsbury's next door, I'm heading in there instead.

There are some Vauxhalls that deserve honourable mentions, though (the Asda 'Extra Special' range if you will). I loved the Astra GTE 16v, with its fantastic 2.0-litre engine, Recaro seats and this cool digital dashboard with 'LCD ELECTRONICS' stamped underneath the speed. That car was better than the Ford of its time. I had one on loan from the Vauxhall Heritage Fleet and spun it because it's so agile that you end up with lift-off over-steer when all the weight moves to the front of the car and the back end goes light and rotates. Once you know how to pull it straight through the right amount of gas, it's great fun. It still felt fast by today's standards because it was light and had that powerful, revvy engine.

When Vauxhall brought out the VX220 in 2000, it surprised everyone. It's another car I've seriously thought about buying and I preferred it to the Lotus Elise, even though Lotus ensured that dynamically their car was just a little bit sharper. But the car only happened because Lotus were low on cash and needed some help. In 2000, Lotus had to replace the first-generation Elise because it failed the new European crash protection regulations. They needed a partner, so GM stepped in to co-fund a new Elise. And in return, Lotus agreed to produce the VX220/Opel Speedster using a modified version of the chassis they designed for the new Elise. At *Auto Express*, we had three VX220s as long-term demo cars and one of the editors crashed the first one. That was the standard 2.2-litre version. We also had the 2.0-litre turbocharged engine and the best of the lot – the limited-edition track-oriented VXR220, produced in 2004. That was some car. Vauxhall had only introduced their high-performance VXR range that year, and it was the first of several cool models, like the VXR8, a crazy, muscular rear-wheel drive coupé. Again, though, it wasn't made by Vauxhall. It was produced by another brand in the GM locker, Holden in Australia.

The Vauxhall Insignia became their flagship model when it was released in 2008. It was really good value and won the European Car of the Year in 2009. I remember advising a mate of mine to get an Insignia. It's a good car, it handles well and it was on sale for a great price. So while I have no real passion for Vauxhalls I do recommend them, when it makes sense. I think they're good, all-round, fairly affordable cars . . . for people who aren't that into cars.

While I've been writing this chapter and thinking about Vauxhalls from the 1980s to today, I keep remembering Vauxhalls that I did really like. The Calibra was beautiful. It had a snappy V6 engine and they did a turbo version that was cool. It's made me wonder if Vauxhall are actually pretty good but they've just been knackered by the high standards of the German carmakers. And because we buy all our cars on PCP (Personal Contract Purchase) or lease them rather than buying them outright with our own cash, a lot of people feel like they'll treat themselves to a higher value car, i.e., not a Vauxhall.

Somehow Vauxhall's cars of the past count for nothing. It's like negotiating with my daughter. I've just given her some chocolate but that was more than a minute ago so it carries no weight any more. It's not fair when you consider that Ford are producing a range of achingly dull cars and yet still the motoring press is interested about their new model. More people seem to be interested in Fords than Vauxhalls and that's reflected in our video-viewing numbers.

I did an amazing trip with Vauxhall, for a car they sold almost none of in the early noughties. Holden, who I've already mentioned, produced a rear-wheel drive muscle car called the Monaro in 2001, and the top-spec GTS could be optioned with a V8 engine. When you looked at the price point of this big, powerful car, it was much cheaper than a BMW M3 of the time. It was a really, really good, fun car, but once again, it wasn't Vauxhall's product. Holden planned to sell it in the UK, so they flew a bunch of journalists out to Melbourne for the launch of this car. We went to the Holden factory and spent three or four days with them, driving the car on their test track and out on the beautiful Great Ocean Road. It was one of my favourite work trips, being flown out business-class to Australia and spending a lot of time with a good car. We visited the Twelve Apostles, the famous limestone stacks in Victoria, driving along beautiful roads in a stunning landscape.

I was quite junior at that point in my career and was paired up with a journalist from a national newspaper, who had a lot of power in the industry. Car manufacturers bent over backwards for him and he was brutal with his reviews, slagging off cars just for lolz. What I didn't know before that trip was that he was someone to avoid due to his sketchy driving. Australia are strict on their speed limits but my god, he did not give a shit. It was frightening. I was too scared to ask him to slow down so I just prayed that I wouldn't end up forming the Thirteenth Apostle in a heap of human and Holden. I was so relieved when it was my turn to drive. 'Why don't you have a little break and put your feet up?' I'd ask, helpfully. Putting aside the *being in fear of my life* thing, it was my first big trip for work, and it was incredible. I think someone else must have pulled out of the trip because I was the lowest of the low there. As a car manufacturer, I would have been annoyed that they sent someone as junior as me on a trip that they'd spent that much money on. I came back and I wrote the two pages that my editor asked me for. It was nuts. Those were the days!

Today's motoring journalists would be expected to have put together four pages on the web on the same evening as they'd driven the car, while simultaneously posting videos of it on TikTok, Instagram, Facebook and X. And to have produced a long-form YouTube video. Along with coming back with four separate news stories and two new ideas for features. Times – and cars – have changed.

Volkswagen

When I was a teenager, a good mate had a Golf Driver, the Mk2 non-injection Golf with the GTI body kit. The difference in quality and performance compared to my Ford Fiesta was like the difference between a pager and a mobile phone. Massive. One of the older kids at my school had a Mk1 VW Scirocco and that was a cool-looking car even though it was twenty years old. I probably would have taken that over a Golf, but not the Mk2 16-valve Golf GTI – the one with the dual headlights, rear spoiler and red, blue and green stripes on the front seats. A top speed of 130mph and 0–60 in around 7.5 seconds sounds like a modern entry-level EV, but back then, that was special. Plus, their marketing was on point, perfectly targeting the male 17–21-year-old demographic. The coolest kids in Walsall had Golf GTIs with BBS alloys, Clifford alarm systems, an Alpine stereo and a Magic Tree air freshener to unsuccessfully conceal the smell of *magic trees* that would often be available from the driver. My uncle actually had a Mk3 Golf GTI but he didn't quite fit the mould – he was a headmaster at a junior school. That's a cool car for a headmaster. You kind of imagine them driving a Volvo 240 or a Rover 800 with walnut dash.

I really started to appreciate VW Golfs when I became a motoring journalist in the noughties. The standard Mk4 Golf, which came out in 1997, was on another level to the Ford and Renault competition. I was hoping to land a Mk4 Golf as my first long-term demo car but that was ambitious for a junior. We were talking entry-level Peugeot, at a stretch.

The new Mk 4 GTI wasn't quite as *wow* as any of the other Golf GTIs, but people still had a lot of affection for them. You hear them spoken about as the OG 'hot hatch' quite a lot on the internet, but it's rubbish. That was most likely the Simca 1100 TI – with its alloy wheels, front and rear spoilers, six headlight arrangement and top speed pushing 100mph with a following wind – which came out in 1974, two years before the Golf GTI. But the Golf GTI was the first hot hatch that was easy to live with, fun to drive and, because it had a fuel-injected engine, it would actually start every time, even in winter. Not that any of this matters – the term 'hot hatch' wasn't even coined for another ten years!

What VW do well is perceived quality. Their cars feel solid, possibly less so now, but certainly in the past. The problem with German cars more widely is that they do sometimes overcomplicate their moving parts. So while things can feel great, they can be so complex that they break and are then tricky to fix when they do. Take the recent debacles with VW's TSI engines and DSGs (direct shift gearboxes).

One VW that I struggled to understand was the New Beetle (released in 1997), which I drove when I was at *Auto Express* in the early noughties. I felt like I was sitting weirdly far back in it. Like a liveried chauffeur driving a three-wheeler, something felt a bit off; slightly comic maybe. The tiny vase positioned next to the steering wheel, designed to hold a single flower stem, didn't make things any better. It looked like something Timmy Mallett would drive. The guy who designed it, J Mays, was later employed as a design consultant on the Pixar movie *Cars* and that made sense to me, seeing as he'd already designed a cartoon car.

The second-generation New Beetle (produced 2011–2019) was much better, partly because the shape was closer to the old Beetle, but I still can't work out what VW were thinking with the New

Beetle. The flower vase made their target demographic pretty clear. That's great if you're developing something with no history, like the Nissan Figaro, which was aimed squarely at thirty- to sixty-year-old female teachers. But if you're going to update the Beetle and try to retain some of the features of the original 'people's car', which appealed to such a wide demographic, why shrink your demographic with a divisive design decision? The New Beetle felt like it was all style over substance. It was less practical than a Golf, more expensive than a Golf and drove worse than a Golf. So get a Golf.

The Golf and the New Beetle shared the same platform – one of the massively significant business decisions made by Ferdinand Piëch, chairman of the VW executive board and grandson of the Porsche founder, Ferdinand Porsche. Under his watch, cars shared the same underpinnings and were then decorated and priced differently. I think of platform engineering like a Victoria sponge – you've got the base and then you can turn it into a birthday cake, a wedding cake, whatever. Stick some sprinkles on and charge what you like.

I remember going to a VW AGM at the company's Wolfsburg headquarters when Piëch was in charge. Even as a little journalist in a corner, I got a sense of how intimidating a character he must have been to work under. If the VW headquarters was engulfed by fire, I can imagine an argument breaking out among the employees over who was going to break the news and wake up the big man. Whichever poor sod it was would have been summoned to the conference room to explain themselves. I can hear 'The Imperial March' from *Star Wars* start to play . . .

Darth Vader ruthlessness aside, under Piëch's leadership, VW became one of the biggest car manufacturers in the world. He played Porsche like a pro poker player, waiting for the right moment and then gobbling them up when a few years before (see Porsche entry for more on this) it looked like Porsche were holding the aces. It's not entirely clear why Piëch left VW in April 2015. It looked like a board power struggle over appointing Martin Winterkorn as the CEO, but the timing did get people wondering, seeing as 'Dieselgate' broke five months later. None of us motoring journalists had any idea it was coming. In fact, every journalist was kicking themselves afterwards for not discovering the 'defeat devices' that could detect when they were being tested and change the car's performance so as to improve the car's emissions. What an amazing exposé that would have been.

The closest I got to the story was a couple of years before when I shared a car with Audi research and development boss Ulrich Hackenberg – one of the senior VW guys who was suspended and later left the company during Dieselgate. Ulrich was explaining to me in forensic detail how adaptive cruise control (ACC) used a radar to both maintain a safe distance from the car in front and keep you in lane. It was the first time I'd driven a car with my hands completely off the steering wheel.

VW had to pay out more than $30 billion in fines and settlements but the bucks didn't stop there. The aftershocks led to cost-cutting that has impacted VW's product quality. And this was all happening while they were trying to create two product lines, their ICE (internal combustion engines) and EV range. VWs have traditionally been well ahead of their competitors in terms of fit and finish and attention to detail but they've both taken a hit post Dieselgate.

First, all the controls started going through the infotainment system to save on the design, labour and parts that go with the buttons and wires you'd need. It would help if the infotainment system was stunningly quick and easy to use, but it feels like they've used a cheap processor because it's laggy and slow. If there hadn't been Dieselgate and electrification, I think there would have been more money allocated to improving something like that in the Golf. It would be a noticeably better car than it is. It's not better than it was eight years ago, and that's a problem.

There were also little details missing, like the illumination for the sliders to operate the climate control settings on the ID.3, although, to be fair, they did fit them after I (and a bunch of other journalists) complained about it. They would never have done something like that on the Mk6 Golf, which was far better than its competitors. A good example of a little detail that VW had included was the design of the parcel shelf. On a regular hatch-back, there are two cords that hang down, fastened by a knot on the underside of the shelf. In the VW, there's a weighted ball which pulls the string taut, which looks neater. The ball's made of a special kind of rubber and it's fitted with these little fins that have a dampening effect, so if you go over a bump or turn sharply, it stops it from banging. They would have spent ages designing each of these tiny features. I always imagined there was a German maths geek drawing equations on a blackboard, who handed the design to an elderly Geppetto figure, who spent all day every day crafting rubber balls. But no longer, because they haven't got the money to invest in this level of product detail.

In the Golf, the trim on the armrest was a bit more plush and it would feature soft touch materials on the top of the door; trims that a Ford Focus wouldn't have had. These little details coalesce to create a more expensive-feeling product. There was nothing in the Golf that annoyed me. If you've seen my videos, you'll know that I put together five good things and five annoying things about cars I review. For VW cars before Dieselgate, I'd be scrambling around trying to find two things that pissed me off, let alone five. And I think that's because they effectively had a department that identified these problems and ironed them out. The Germans have probably even invented a word to precisely describe exactly that activity, the way they did with *Schatten-parker* (one who parks in the shade).

On the ID.3 (basically the electric version of the Golf), rather than having switches or buttons for all the windows on the driver's side, they've got buttons for the front right and left windows. Underneath is another button, which you press to toggle the mode so the window buttons activate the back windows. Some things in a car you need to work immediately. The door locks and windows, for example, because your kid will open that window and that door at some point. In that moment, I don't want to be looking down and trying to work out which button it is and which mode I'm on. Maybe none of the designers had children, I don't know, but what they've come up with is utter bollocks.

I think this is all part of this post-Dieselgate cost-slashing. VW might be the largest car manufacturer in the world in terms of revenue but they're facing challenges. They were planning to close three factories in Germany in 2024 and asked their workers to accept a 10 per cent pay cut. After negotiations with the trade unions, they abandoned the closures but are still committed to cutting more than 35,000 jobs in the next five years in Germany to save money. But it's not all doom and gloom in Deutschland. At VW in 2023, they sold 8.5 million of part number 199 398 500 A.

But it's a sausage.

That's right, VW produce their own currywurst and there's an accompanying thick tomato ketchup with part number 199 398 500 B. The currywurst is made at their on-site butchers and sold in the thirty-odd staff canteens across the VW factories and in the wider world. When VW decided they planned to give the sausage the chop in 2021, former German chancellor Gerhard Schröder mounted a passionate defence of the sausage, which he referred to as 'the power bar of the skilled factory worker'. Just keep clear of the exhaust.

VW are actually one of my favourite brands, and I really hope they turn it around. They're good at evolving and adapting. For example, the Mk7 Golf R (released in 2014) had a good engine, a decent gearbox and good traction from its four-wheel drive but it was seen as a bit of a one-trick pony, so VW took on board the criticisms and feedback from motoring journalists and the public. They kept all that was good about the Mk7 Golf R and then added in the fun element that was missing for the much-improved Mk8 Golf R in 2021, including a clever rear differential, which means the car is able to drift, and that makes it so much more playful.

When VW get it right, they absolutely nail it, and the Golf R Mk8, recently facelifted in 2024, is a good example of that. Another is the comfy, spacious and easy to drive Tiguan – VW's bestselling car. Their tech is a little bit behind but it's a good, solid, all-round car. And that's what the company was built on. Volkswagen – the people's car. VW aren't necessarily the best manufacturer in any single category, but they score well on every-thing. They're like the school captain of cars. They're dependable, score well on paper, pretty sporty and decent to spend time with. They're popular with cool kids, dweebs, parents and teachers alike. That's Volkswagen to me.

Volvo

I've had a few Volvos and I've liked each one. The high-performance version of the C30 D5, with its five-cylinder 2.4-litre engine, was both smart and pretty fast. I was quite young at this point – well, in my thirties – and I'd had this car for six months, so I was just about to put together my assessment/end-of-term report about it. It was at this moment that I realised I hadn't floored it, kicked down the gearbox or thrashed it around at all. I was in a high-performance car and hadn't actually experienced the performance. Here's what I realised: there's something about a Volvo that turns you into a calm, sensible driver. I had to make a conscious effort to drive it fast but it didn't feel right. A Volvo's like a gentle pleasure cruise, bobbing along looking at fjords. Everything's kind of minimalist: clean and crisp and relaxing. You go into a Scandi chill meditative mindset. Maybe we all need to be driving Volvos.

I had a C70, with a folding metal roof, and I soon discovered the main problem with this feature. I was doing a photo shoot and we wanted to catch the moment the roof was halfway through the folding process, but the photographer kept on missing the shot, so I had to do it about six times. The seventh time, the roof forgot where it was in the process. So when I finally shut it, it came in too low and smacked into the windscreen pillar, which put it out of whack. It was still under warranty and I took it to a dealer, who had the car for a month to fix this roof. If it'd been out of warranty, I probably would have written it off because the hours spent fixing this roof were longer than the hours required to build the roof. There are some features I miss about cars. The folding metal roof fad ain't one of them, and that's a rare subject that unites all car owners.

Volvos are usually front-wheel drive or four-wheel drive, but the 340, which came out in the late 1970s, was a rear-wheel drive hatchback. It's an ugly car and I hated it as a kid. But they experienced a new lease of life in the early noughties, because you could pick them up extra cheap and they proved to be perfect for learning how to drift. I also came round to the shape. I've come to think of Volvo 340s like coffee – you can't stand it when you're young, but your tastebuds change over time and suddenly you realise what the fuss is about. Maybe that was the Volvo plan back in the 1980s and 1990s: *you'll get it, eventually.*

As is often the case, the Swedes knew what they were doing. I can imagine there were howls of laughter when a Volvo estate turned up at the British Touring Car Championship (BTCC) in 1994. Volvo had kept it a secret until the day. But this was no ordinary estate: this was the 850 T5-R. The coolest Volvo ever made.

Four years later, a Volvo 850 T5-R won the BTCC. It was also winning in the consumer market. With its 2.5-litre, five-cylinder turbo generating 230bhp, the 850 T5-R was the fastest estate on the market, and that fact wasn't lost on police forces and cool dads all around the UK. Another cool Volvo estate was the V90 Cross Country D5, which Volvo invited me over to test in 2018, in the Arctic Circle. I spent the first day driving on roads, which ranged from lots of asphalt, to a suggestion of asphalt, to lots of ice. The second day involved driving on a frozen lake in a kind of enchanted wonderland. Imagine the dance scene towards the end of *The Snowman* just before they meet Santa and you've got the picture. Sliding around on that track felt like Christmas had come early.

Like the words 'IKEA' and 'flatpack' go together, so do 'Volvo' and 'estate'. But in the summer of 2023, the unthinkable happened. They announced they'd no longer be making saloons and estates for the UK market. It was like IKEA announcing that meatballs were off the menu; people were upset, but then they remembered that they never bought them. Volvo had become an exclusively SUV brand well until they reintroduced the Saloon in 2025.

Aside from their estates in the 1980s and 1990s, Volvos were kind of also-rans. They made decent cars, sure, but no one was going to want their saloons over Mercedes or BMW back then. Maybe that was the point, though. It was almost like they weren't directly competing with other brands but occupying the space in between. It was the car that you went for if you were looking for a little upgrade from a Ford. Or if you wanted to pay a bit less than a BMW and were after something a little larger. A bit like choosing a slightly less desirable house to give you more room and a few more quid in your pocket. That was the kind of game they were playing for a while, and they did all right. But in 2014, they tried something different. They made a car that people actually wanted: the second-generation XC90.

The second-gen XC90 was a seven-seater SUV that looked great, felt expensive inside and really was desirable. And from that came the second-gen XC60 and the XC40. I had an XC60 with a beautiful cream interior as a long-term demo and it was a great car. Suddenly Volvo were leading the way in the SUV field. They'd managed to turn the brand around from *fusty old man* to *Scandi cool*. And they've done all right in the transition to electric.

Volvo make good electric cars partly because they're owned by Chinese company Geely, which gives them decent battery tech. The EX30 is the smallest SUV in the Volvo showroom, but it's such a good model. It's stylish, really comfortable, affordable and quick. Their Twin Motor version gives you all-wheel drive but also makes it super-quick. At 3.6 seconds for 0–60mph, it's the fastest-accelerating Volvo ever made. It's also one of the fastest around at the price point we're talking. Put all this together and you can see why it won Carwow Car of the Year in 2024. They smashed it out of the park with the EX30. It helped Volvo achieve their highest full-year global sales since they were founded in 1927. Who'd have thought that a car brand, who named themselves after the Latin for 'I roll', would know what they were doing?

Well, it turns out that 'I roll' is something Volvo take very literally. I was invited out to Gothenburg about twenty years ago to watch the Volvo guys deliberately roll the first-generation XC90 (the one before it started looking cool) to show how strong the body structure was. So they brought out this sled, carrying an XC90, which is at a right angle to the direction of the sled. They sped up the sled to like 50mph and then stopped it dead, so the car immediately rolled over about a million times and landed upside down. A bulldozer appeared and righted the car. Then the Volvo guys beckoned me over to the car. I reached for the handle, not imagining for a second that I'd be able to open it. It opened and closed just fine. Absolutely nuts. According to the UK government's crash records, which have been going since 2004, not a single person has died in the UK in a car-to-car accident in an XC90 as of May 2025. Volvo's commitment to safety is something else.

BEST FAST ESTATE CARS OF ALL TIME

1. **Audi RS 6 GT** – The RS 6 is already a brilliant car but the limited-run GT model turns things up to eleven with superior handling and even more poke from the engine.

2. **BMW M5 Touring (E60)** – This is the BMW M5 fitted with a V10 engine. They can be problematic but when they work, my god do they work. Like having a Formula 1 car that you can take to the tip.

3. **Mercedes E 63 Estate (S212)** – Great for carrying your dog about. The S212 is my pick because it has the fabulous, naturally aspirated V8, which sounds like a Messerschmitt Bf 109 fighter plane.

4. **VW Golf R Estate** – Fun to drive but immensely practical. An easy sell to your wife!

5. **Subaru WRX STI Wagon** – A rally car for the road with enough space to carry a spare set of tyres after you've burned through the other ones doing four-wheel drive drifts all over the place.

6. **BMW M3 CS Touring (G81)** – Whenever BMW do special versions of their M products, they are always a sure-fire hit and this is the car that does it all. Super-fast, fun to drive and practical.

7. **Volvo 850 T5-R** – This car was made cool by its success in the British Touring Car Championship and there's a good reason why the police used them as interceptors. They're highly practical and bloody fast.

8. **Porsche Taycan Sport Turismo** – I had to put an electric car in somewhere and this is the one. It's fast, you can get it relatively cheaply as a company car and it's practical-ish.

9. **Citroën CX Turbo Estate** – Punchy turbocharged engine combined with an aerodynamic shape, avant-garde interior and clever hydropneumatic suspension which was so good Rolls-Royce licensed it.

10. **Skoda Octavia vRS Estate** – With this car, you get the engine from the Golf GTI with a more practical, better-value-for-money body. Win-win.

Zeekr

I'm thankful to Zeekr for stopping this A–Z of cars from ending on a V, which would have been a bit pants.

Zeekr are a Chinese luxury electric car brand, which is owned by the multinational conglomerate Geely. The chances are that you might not have heard of these guys, but much like a suspiciously quiet Monopoly player, they've been buying up most of the board without anyone noticing. To give you an idea, Geely own 78.7 per cent of Volvo, 51 per cent of Lotus, 50 per cent of Smart, 49.9 per cent of Proton, 22 per cent of Polestar and 17 per cent of Aston Martin. They also own 100 per cent of the London EV Company, who make London's new electric black cabs.

The first time you encountered Geely may have been for their 'versions' of well-known European, Japanese and South Korean cars in the late 2000s. The Geely Merrie 300 was the one that looked like a low-res Mercedes C-Class. And then there was the 2009 Geely GE concept car, which, let's say, 'drew inspiration' from the Rolls-Royce Phantom. While the inside looked more than a little familiar, it also contained unique features such as a throne-like single rear seat – ideal for the solo businessman or travelling dictator. While we were all laughing at the 'look at these hopeless car copies' articles online, China was busy becoming the world leader in tech, building space stations, hypersonic missiles and robotic taxis. Oh, and making giant leaps in battery tech. The vast majority of the high-end European car makers aren't making their own batteries – they're getting them from China.

I haven't driven a Zeekr but I know that they're already popular in Europe and some of the cars look pretty good, like their Zeekr X SUV. It's aimed at the Tesla Model Y customer, only it costs a fraction of the price, and I'm not sure if you're getting any less. The interior quality is strong, the battery's good, and the tech on the inside is excellent. All of which isn't great news for Elon.

As for the name 'Zeekr', they tell us on their website that Ze 'stands for Zero, the starting point of infinite possibilities', the 'E' is for 'Evolving the Electric Era', and 'Kr' is for 'Krypton, the legendary weakness of the all-American hero, Superman'. OK, so I made that last bit up. They went with Krypton because it's 'a rare gas that emits light when electrified'.

Zeekr have designs on becoming a global brand and they're really not hanging about. They hit the ground running with industry legend Stefan Sielaff on board from the brand's inception. He's the guy who designed the Audi A1 and A7 before he took up the role of design director at Bentley, overseeing the third-gen Bentley Continental GT and Flying Spur. Zeekr have built a state-of-the-art global design centre in Gothenburg, Sweden, and everything's happening at light speed. The Zeekr 007 went from design sketches to finished product in two years (that's about half the time it ordinarily takes). Zeekr have already got eight cars in their range. They have the tech, the talent and the cash to splash. And they've just announced they'll be launching in the UK in 2026.

I've got a feeling that when it comes to updating this book, my list of car brands is going to look rather different . . .

Acknowledgements

First off, massive thanks to Jack – without your push, this book probably wouldn't exist. You gave me the impetus, and now here we are, tens of thousands of words later.

To Ben, for your unwavering belief, Zen-like patience, and ability to nod encouragingly even when I was waffling. And to the brilliant team at Penguin, thank you for your support, guidance and patience.

A big shout-out to Nathan, whose constant input was not only invaluable but also saved this book from going completely off the rails. And to Richard – mentor, sounding board and occasional therapist – thank you for helping me navigate the chaos.

I'm grateful to the entire Carwow team for their flexibility, and to James, our founder, for giving me what might just be the best job in the world: getting paid for driving cars and saying things about them!

Thanks also to the motor industry itself – for being weird, wonderful and constantly evolving. You've kept me fascinated and fuelled with content for years.

Finally, a heartfelt thanks to my mum, who taught me to write properly even after my teachers had written me off. And to Jo, my partner: thank you for the time, the space, and for tolerating endless car chat. This turned out to be a slightly larger project than expected. Sorry about that.